REVEALING THE DEEPEST SECRETS OF KABBALAH

SHIMON ELIEZER "THE S.E.G."

To order additional copies of this book, contact
Toll Free +65 3165 7531 (Singapore)
Toll Free +60 3 3099 4412 (Malaysia)
www.partridgepublishing.com/singapore
orders.singapore@partridgepublishing.com

ISBN
ISBN: 978-1-5437-6960-9 (sc)
ISBN: 978-1-5437-6962-3 (hc)
ISBN: 978-1-5437-6961-6 (e)

Print information available on the last page.

04/12/2022

PARTRIDGE

DEDICATED

TO MY

LOVELY

DAUGHTERS

NATHALIE & LINDA G.

AND TO THE MEMORY OF MY BELOVED FATHER

INTRODUCTION

What is kabbalah קבלה ?. Kabbalah in hebrew means reception. In other words the reception of high,from the Creator Himself.

To be able to comprehend the Esoteric world,you have to be totally open minded and receptive to this voyage into the world of mistery,miracles.

Science fiction is only a kids game in this secret teachings,that will reveal the Unknown and to the answers of questions that you never dared to ask.

Also words that you have heard priviously,as let say ARMAGUEDON,all know the meaning that will be the end of times,but what few know,that it comes from the Hebrew words of HAR MEGUIDO,הר מגידו the mountain of Meguido that is mentioned in the profecies of EZEQUIEL,this Mountain is in the north of Israel,and it will be the war of Gog and Magog גוג ומגוג(to read by same Author)

Another example of un-understood profecies that the messiah will arrive mounted on a white mule. The real meaning is that we will be mounted on our evil habits or negative impulses,that we are stronger by being able to totally control them,only then the messiah will arrive. The mule signifies the "hard-headed" as we are,we know of many things that are wrong or bad for our health or well being but we are still doing them.

We are all connected to .CHOCHMA-ILAAH. חכמה עילאה Meaning Intelligence from high. We are connected, like a computer to a super computer,by a modem.

At no moment I am implying any disrespect to our CREATOR,but to the contrary,it is as an example.

Kabbalah has been the secrets of secrets,for thousands of years .It use to be reserved for the sages,our forefathers. Mystical tales and secrecy kept outsiders as well as insiders,even today rabbis and others,believe in the existance of the Kabbalah,the esoteric hidden secrets,but are afraid to come close because they are afraid of the unknown.

Rumours that people that entered the PA.R.DE.S פרדסWhere after a while losing their mind.PARDES in hebrew means Orchard. It is composed of four letters,PE(P) פ -RESH(R) ר-DALET(D) ד-SAMECH(S) ס. In other words

PE-פ=PSHATפשט(literally)

RESH ר=REMEZ(clues)

DALET ד=DRASH דרש(homilitic)

SAMECH ס=SOD סוד(secrets).

so,entering the Orchard,was entering the world of secrets.

This is the four dimensions that we divide our studies in the depths of Esoteric teachings.

As you will study this very special book,you will understand,first why people believed that most of the persons learning Kabbalah were becoming crazy. In those times we did not have todays technology to use as an example, because Kabbalistic terminology does not have a dictionnary to tranlates the words,intead we can compare to things or parables.

We have entered the era of AQUARIUS, the MESSIANIC ERAביאת המשיח , where we are supposed to learn the secrets of life,of our world and the universe and on top of all, our CREATOR.הבורא עולם

We are in the last day of creation,there are no more new souls,meaning time to correct our soul,because we have the possibility to reincarnate four times for TIKUN תיקון(correction).

Our soul(or Energy-Intelligence) given by the CREATOR From his own ESSENCE. From this essence exists all Creation

The Bible תורה is a testament for all times. It Has 70 facets,different levels,from its simple textual historical facts,to the deepest secrets of creation,and all true. King Salomon (KOHELET קהלת)said:"There is nothing new under the sun".

The bible is coded and all our history that has happened and that will happen is inscribed in the TORAH תורה. By translating from LASHON HAKODESH לשון הקודש (sacred-tongue) HEBREW we loose the esoteric messages.

The letters are cosmic Symbols that have a lot to reveal to us,the vowels and encantations,all the above have the secrets of our passed and future.

I will tell you a small story,there was once two friends,one a beleiver in higher authority,and the other a non beleiver. One day the non beleiver came to his friends House and saw that he had a fantastic robot that looked Almost HUMAN and he said:" it is a beauty,It looks so real,who made it? "The beleiver sade "NO ONE,IT MADE ITSELF " . Oh come on reply the other how can it be made by itself?" Our friend answred why for this thing so simple you do not believe that it was made by himself and the universe that is so complex you say that it was created by ITSELFand not by a higher being,the CREATOR?. Okey said our friend,how does it work?

The answer was"read the instructions", in other words if you want to understand creation read this book"THE KABBALAH".

My main intentention for writing this book,is the task to impart,to teach and help my fellow Human Beings.

If only I secceeded in helping one person,my gole has been reached.

I do hope that this book will enlighten the rest of your life,and the darkness that still exists in our time.

The S.E.G ה שאג.

PART ONE

RECOMMENDED ONLY FOR:

SCIENTISTS, DOCTORS,

KABBALISTS

&

ASTRONOMISTS

The Bridge Beetween

Metaphysics - Physics

Astronomy – Anatomy

&

Religion

THEORY : B

הזוהר- רבי שמעון בר יוחי

THE ZOHAR (book of splendor) Rabbi Shimon Bar Yochai

(A.C 120)

"גוף האדם הוא המפה של האוניברס"

"Man is the blueprint of the universe"

"העולם הוא לא מקום לבורא אבל הבורא הוא מקום לעולם"

"The world is not G-D's place, but
G-D is the place of the world"

הוא נקרא מקום,ספר יצירה

"He is called MAKOM(place)"- SEFFER YETZIRA

THE BOOK OF CREATION.

CHAPTER I

"The man, is a blueprint of the Universe". Therefore, the Universe is a blueprint of man. Let us expand on the theory that will lead us to Genesis, and the sub-atomic world. I will explain the clues given to us in the Bible, the most authentic document of all times. I developed this theory through the above mentioned phrase, and of course from Kabbalistic information.

I have interpreted the entire first chapter of Genesis in the beginning of the book. Now let us discuss "The Blueprint of Man". Adam, as I mentioned previously, came from the word "ADAMA" (earth). Let's analyze this specific passage at face value, that talks about the earth and the entire galaxy.

We will have to go alternatively to man's anatomy and galactic compositions. (Genesis 1 :26) G-d said, "Let us make man with our image and likeness". Likeness of the universe.

The sub-atomic world would yield clues and similarities. Man is made of cells. And each cell of atoms. Atoms are composed of protons, neutrons, electrons, etc. I am going to write a general physics information on a very specific topic.

At the turn of the century (twentieth), physicists discovered that atoms themselves were made from smaller pieces; i.e. sub-atomic particles. These smaller pieces first appeared to be the ultimate limit of matter, but in the 1960's, physicists proposed that many subatomic particles were themselves made from smaller particles that could not be detected in the usual manner. Like the unobserved atoms of the 19th century, the undetectable smaller particles, called quarks, have come to be accepted. Possibly in the next century, there will be "photographs" of quarks.

Quarks are considered to be "fundamental", scientists believe they are not made up of smaller pieces. Besides the quarks, various other groups of particles are thought to be fundamental.

Among these are the "Leptons". Together, quarks and leptons form what we believe to be "matter". They are characterized by a spin of 1/2, as are the particles made from three quarks, the Baryons. Spin is a number for each particle that has behavior similar to "Angular Momentum".

Other apparently fundamental particles are similar to the Photon, the particles that make up light. Particles such as photons and particles with a spin of either 0 or 1 are called Bosson, (which act as

glue) that hold matter together. Bossons produce the four known forces of the universe; gravity, electromagnetism, the strong force and the weak force, which become a single force, the electro-weak force.

Many subatomic particles carry a charge that is a unit of electromagnetism. Nearly all charges are counted as, either –1 or +1 or 0 (no charge) based on the change of the electron which is -1. Quarks however, have charges in multiples of 1/3rd. No one knows why charges come only in these particular amounts and do not occur in other amounts. No known law, for example, predicts that the charge of the large Proton will be exactly the same amount (but opposite: + 1) as the charge of the small electron.

The masses of the subatomic particles are measured in terms of particles energy. Energy (E) is related to mass (M) and the speed of light (c) by Albert Einstein's famous E=mc2. Since c is a very large number, the energy of a particle is much larger than it's mass. Even so, the energy is expressed in a very small unit, the MeV which is a million volts.

Recently many physicists have proposed other still undetected particles called wimps (for 'weakly interacting massive particle'). These are far from being fully accepted at this time and are omitted from the following list.

Every particle mentioned has an anti-particle whose charge is opposite (negative particles have positive anti-particles) and, whose spin is in the opposite direction. Unless there is something special about the anti-particle, it will not be mentioned.

Reviewing the main components of the atom:

Fermions - All spin ½ Fundamental particles of matter.

The Electron (a Lepton): Movement of electrons is a source of electrical current, while an excessive deficit of electrons causes static electricity. Electrons are found in all atoms: Where they occupy several shells around the outside of the atoms Like the solar system and the stars around the galaxy, the sun being the proton and the stars around it, the electrons. The galaxy just being an atom, the Earth a quark.

The electron is stable and very light, like all subatomic particles. It has a related particle.

The anti-particle is known as the positron. The positron is a mirror image of the electron.

The electron is a -1, the positron a + 1.

The positron was the first anti-particle discovered.

Muons and Tauons (Leptons): The Muon is often described as "fat electrons" since it has all the properties of an electron except that it's mass is 200 times as great.

Similarly, the Tauon is a fat Muon (12 times bigger). No one predicted these particles and so no one knows what their role in the universe is. The Neutrino family (Leptons): is a group of leptons associated with electrons, muons and tauons with which they form three "families".

A family consists of charged particle, such as an electron. It's anti-particle such as a positron and an associated neutrinos and anti-neutrinos. They do not interact strongly with anything.

Neutrinos are passing through your body and the earth all the time. The Baryon is composed of three quarks. Common Baryons and Mesons are composed of quarks known as up and down. Even though their individual masses are small, the binding energy between them is a baryon which produces most of the mass of the universe.

Other Baryons and Mesons have a quality known as "strangeness" which is composed by the strange quark or a quality known as "charm" composed by the charm quark.

Two other quarks variously known as top and bottom or as truth and beauty complete the six flavors (like the six components that I explained in the Power of letters). Furthermore, quarks are confined with the particles they make up, making direct detection impossible.

Baryons

The proton is found in the nucleus of the atom. The proton appears to be stable. Protons are heavy particles where every atom contains an equal number of + I protons and -I electrons, giving a total charge of the atom of O. An atom that has lost or gained an electron (-) and therefor a charge is called an ion.

Neutrons are almost exactly like a proton, but with no charge therefore much harder to detect. Neutrons in the nuclei of atoms are stable, but outside the nucleus each neutron soon decays into a proton, an electron, and an electron anti-neutrino. All atoms except hydrogen must have neutrons in their nucleus in order to be stable.

Other Baryons: These include two hyperons, three sigmas, two xis and an omega. None of these particles are stable, decaying after much less than a second into other particles; i.e. they do not exist at ordinary energies on earth.

Gluons: The quarks theory soon led to the understanding that pions are the strong force and side effects of a more essential strong force, one carried by the eight neutral particles called :

Gluons. Exchanging gluons between quarks usually causes quarks to change from one color to another, keeping the quarks attracted to each other.

Two Basics Laws of Quantum Physics

When one considers the effects on very small masses and at very small distances, different forces begin to affect how objects behave and occur at the sizes and distances one can observe directly. Since these effects occur in discreet steps, as with Plank's Quantum, which is the size by which energy changes in steps (instead of continuously). The science of such effects is called quantum physics. Small masses sometimes act like particles and sometimes like waves and sometimes like nothing we know about at the scale we live in. Two laws that describe their behavior in particular are basic and easily stated.

1. Heisenberg's Uncertainty Principle: It is impossible to specify completely the position and momentum of a particle such as an electron.
2. Pauli's Exclusion Principle: Two particles of such a class that are essentially the same cannot be in the same exact state. This class, the fermions, include such particles as the electron, neutron and proton.

Particles of different class, the bosons, do not obey Pauli's Exclusion Principle.

Principle Elements of Astronomy

Binary Stars: Almost half of the stars in the visible universe are actually pairs of stars that orbit each other. Astronomers can sometimes see both stars, but more commonly they recognize that a star is part of a binary, because of the influence of the dimmer star's gravitational pull on the other star.

Black Holes: When a body becomes so massive for it's size that not even light can escape the powerful gravitational pull it exerts, it is called a black hole.

Brown Dwarfs: Are bodies, too small to be stars (since they emit no visible light and are not undergoing fusion), and too large to be planets (they give off a lot of energy in the infrared part of the electromagnetic spectrum, as a result of gravitational contraction). No brown dwarfs have been positively identified, but some astronomers think there might be enough of them to account for the "missing mass".

Expanding Universe: The farther away something is from us, the faster it is moving away from us. When Albert Einstein developed his general theory of relativity, he found it predicted that the universe would expand as if it were exploding. In the expanding universe, everything is moving away from everything else.

Galaxies are systems of many stars separated from one another by largely empty spaces. Great masses of stars become known as galaxies, after our own Milky Way, the galaxy which includes the sun. There are two types of galaxies - spiral and elliptical, although some are neither (irregular).

Milky Way: This is the galaxy to which the Sun and the Earth belong. Early 19th century, William Herschel determined that our sun was a star in a vast lens-shaped star system, and that the Milky Way was the part of the star system.

Missing Matter: is matter that is apparently in the universe but that has not been discovered or observed. Astronomers note that galaxies are rotating as if there were bodies embedded in invisible layers. Furthermore, there are theoretical reasons to believe surrounding galaxies there is even more matter in the universe than can be accounted for by the invisible matter.

The surrounding light of the EIN-SOF (The Creator's Essence). Ideas as to what this "missing matter" might be included everything from brown dwarfs to undiscovered subatomic particles.

Nebulae: Are patches of gas and dust both of which are observable in telescopes. It was discovered that some cloudy patches were past vast collections of stars, all such patches were called Nebulae (meaning clouds). Some are galaxies, some are not. These patches of gas emit light. Some patches of dust also glow, usually reflecting the light of nearby stars. Some nebulae have the shape of North America - or the Horse's Head nebulae, another class of nebulae looks similar to giant spheres.

Novae: are stars that seem to appear in place of dim stars out of nowhere. They don't just appear, but instead, a dim existing star, suddenly brightens.

Neutron stars: are those composed of neutrons instead of atoms. When a star's core collapses, it may collapse enough so that the electrons and protons in the core are squeezed together to become neutrons. Such a star may be only a dozen miles in diameter but may be a mass twice of the sun.

Pulsars: are neutron stars that emit radio signals from their pole in a direction that reaches Earth. All neutron stars emit radio signals from their poles and rotate very rapidly. These signals form a tight beam. If the beam intersects Earth, we can observe through a radio telescope, a fast pulsing on and off, like a signal. The pulses are so regular that when they were first discovered, they were thought to be the work of extraterrestrial beings.

Quasars: Are distant sources of great energy. The name quasar, is short for Quasi stellar object. The objects are so called, because they seem to be about the size and general appearance of stars, but produce far too much energy to be stars. No one knows for sure what they are, but there is some evidence that quasars are the central part of distant galaxies. The stars in the galaxies cannot be seen because of the great distance so we only see the central part, which is the quasar.

Red Giants: Are the stars that have used their hydrogen fuel and expanded as a result Young stars combine hydrogen in a nuclear fusion process that forms helium. When a star has consumed the hydrogen in it's core, new fusion reactions that starts with helium begin, leading to carbon.

Stars: Are bodies of gas large enough to undergo fusion reactions in their core. As a result of the energy produced by fusion, stars emit visible light as well as electromagnetic radiation at other wavelengths. The sun is considered a star.

Superclusters and Clusters of Galaxies: are groups of galaxies associated in space. There may be just a few members of a cluster or as many as thousands. About two dozen galaxies near us form with the Milky Way, our Local Group. The members of the local group also include the Andromeda Galaxy and the Large and Small Magellanic Clouds. All are traveling through the universe together. The Local Group is a member of a supercluster of galaxies, called the Local Supercluster, that contains about 100 clusters. Clusters and superclusters are primarily recognized because the average distance within a cluster or supercluster from one galaxy or cluster to another is much less than the distance to other galaxies or clusters.

Supernovae: Are large stars that explode. A supernovae explosion is much more drastic than the brightening of a Novae. A supernovae reported by the Chinese astronomers from A.D. I 054 was visible in the daytime. The remnants of this explosion are known as the Crab Nebulae. At it's heart, the Crab Nebulae has a pulsar, all that is left of the star that exploded.

Variable Stars: Are stars that periodically change brightness. Some variables are part of a binary system in which one star periodically passes in front of the other.

White Dwarfs: Are stars whose core have collapsed until all the atoms are pressed very close together. The core collapses because a red giant has used all it's helium for fuel, but the star is too small to start burning it's carbon.

Human Anatomy

The Body:

There are 9 main systems that the human body consists of:

The skeletal system

The musculature system

The nervous system

The hormonal system

The circulatory system

The digestive system

The respiratory system

The immune system

The reproductive system

Cells are composed of the nucleus, cytoplasm, cell membrane, and various smaller parts that have different functions.

Up to now I have cited physics, Astronomy, and Human anatomy. Now I want to show the common points which are in the three. Ultimately I intend to demonstrate that we are a blueprint of the universe. If we explore the human body and understand it fully, we will know all the secrets of the universe.

Meaning:

G=nth U=nth

Galaxia=Atom Universe=Body

As the universe is composed of: galaxies and nebulas and supercluster.

The human body is composed of: organs, cells and ultimately, atoms.

The universe has the form of a human being. I mentioned the four worlds, EMANATION, CREATION, FORMATION, AND DEED, they are 4 dimensions. The body has 4 dimensions, (See figure 1),

1) Skeletal cells,

2) Muscle cells,

3) Blood cells,

4) Skin cells.

From within, all are very similar. Following the theory that man is a blueprint of the universe. We are in the Milky Way galaxy, together with other galaxies, forms a cluster, or cell. This cell, formed of a few galaxies, is equal to a human cell that is made of atoms. What I am saying, is that the galaxy is an atom; and a nebulae or supercluster is a cell.

Our galaxy as an atom:

Our Galaxy = An Atom

The Sun = Proton

6) Jupiter = hyperon

7) Saturn = hyperon

8) Uranus = hyperon

9) Neptune = sigma

4) Earth = omega-Quark

3) Venus = xis

5) Mars = sigmas

2) Mercury = Neutron

10) Pluto = xis

The Moon of Saturn = Gluons

Other unseen stars = Pions

Stars in General = Muons, Tauons, Leptons

Meteors, Asteroids = Neutrino (that goes through a human body and we do not see it, similarly are the asteroids, they disappear.)

All of the above = Galaxy = Atom.

A group of galaxies forms a nebulae or cluster. A cluster of atoms, is a human cell. (See figures No. 16, 17, 18, 19, 20, 21). With so many similarities, we can assume the theory, that the human body, is a blueprint of the universe.

FIGURE No.16: Example of Atoms & the Solar System.
The similarities of the Atoms with our Solar System.

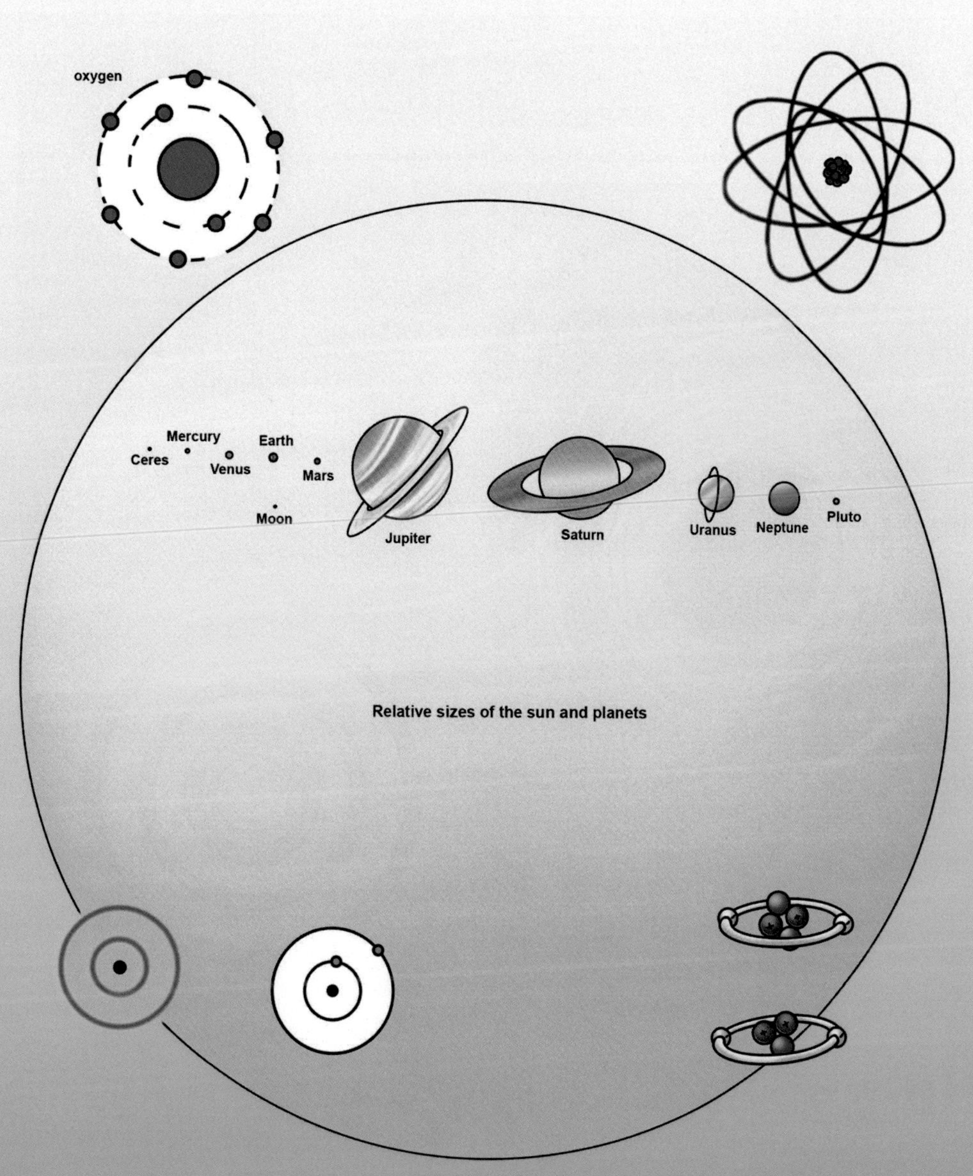

FIGURE No.17

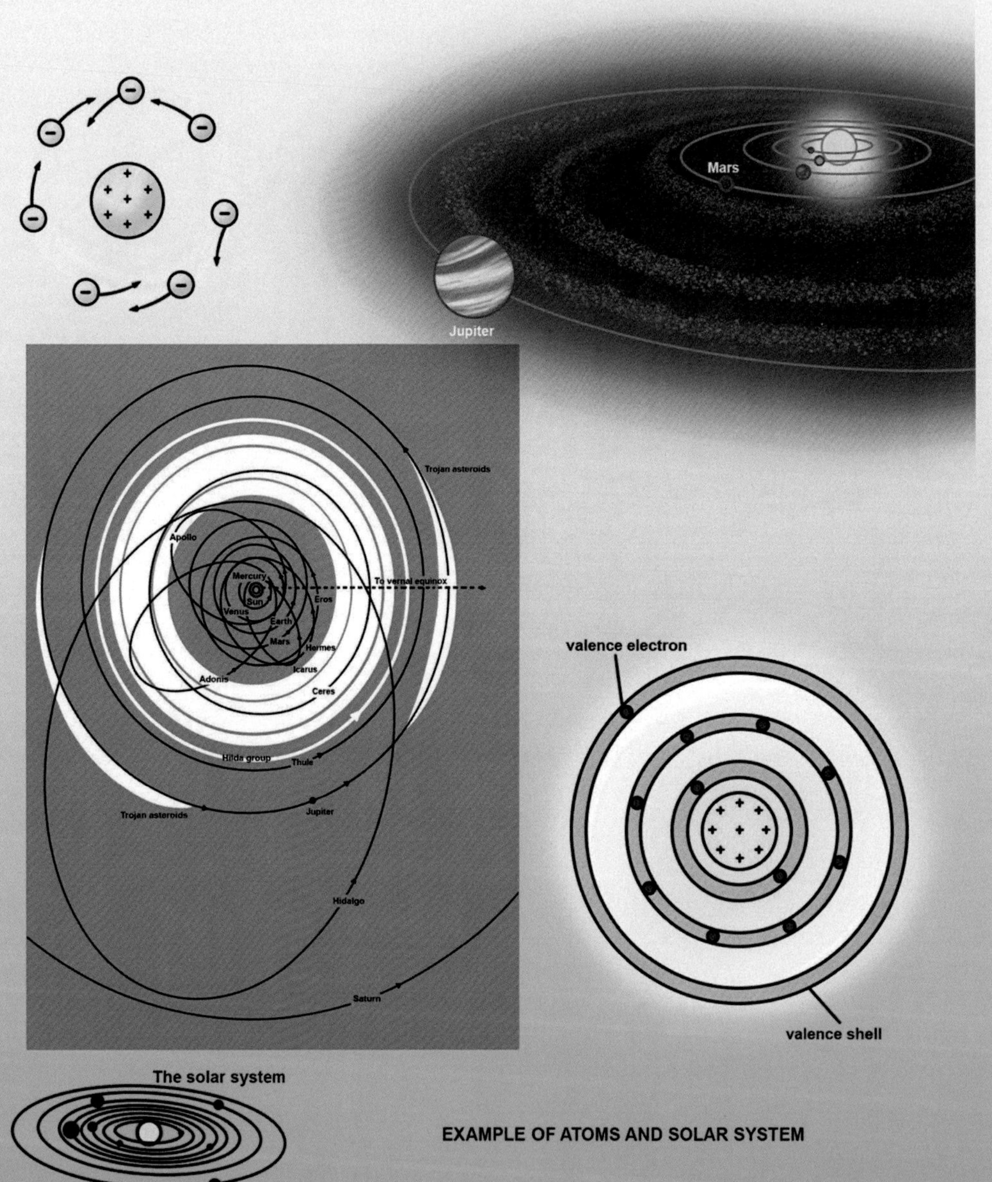

EXAMPLE OF ATOMS AND SOLAR SYSTEM

FIGURE No.18: Our Galaxy & Others The Atom we belong to.

FIGURE No.19: A Bone Section,A Galaxy & A Bone Cell.

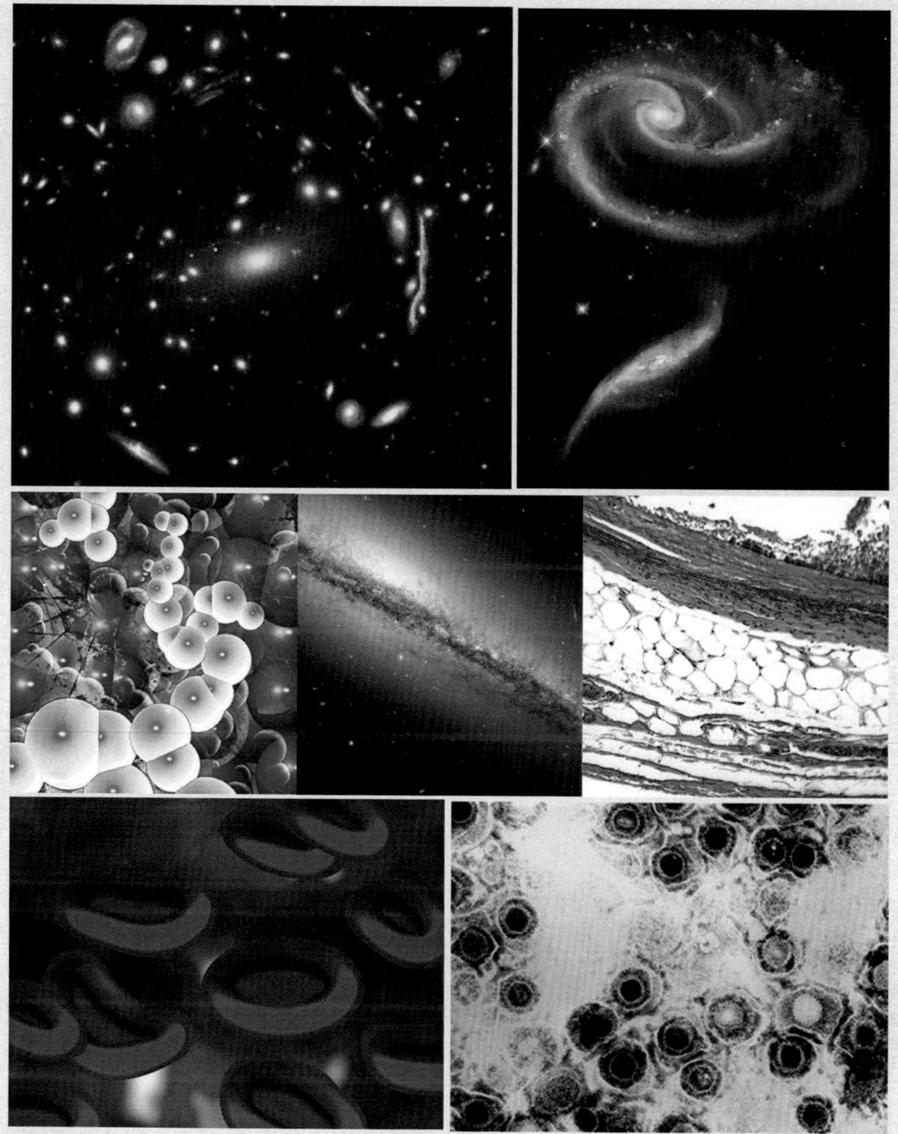

FIGURE No20: Example of a Human Cell & a Supercluster.Here we can appreciate the similarities. The top pictures show a group of Galaxies and below Terrestrial cells. Below a Peculiar Galaxy and beside it a Human Muscle Cell. As I said before, this will be the equivalent of the world of FORMATION.

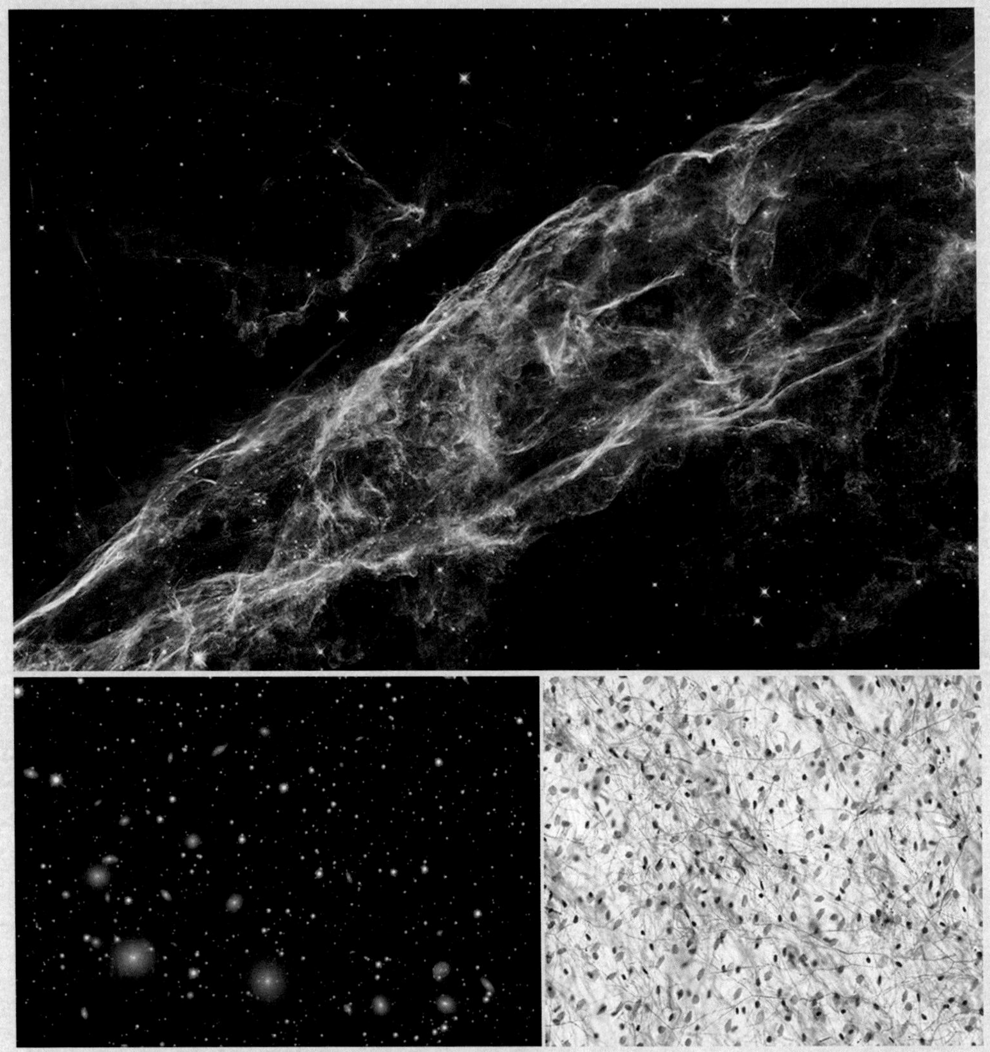

FIGURE No.21: Super-Clusters & Skin Cells.

I was not able to find real similarities of Galaxies,to skin cells,it is because the Skin is equivalent to the end of the Universe or ATSILUT,EMANATION. The Unknowable,Our CREATOR,what MOSES wanted to see and was turned down by the CREATOR.

If we are so, our galaxy is an atom of a cell of a colossal entity but colossal is just a word, an expression, because there is no word that could express what I am trying to say. I am bringing a quote from the Book of Raziel, the Angel,the most secretive book of Jewish mystics.

Rabbi Yishmael said: I've seen the King of Kings sitting on High and His armies standing by Him on His right and on His left, (and the Angel, Minister of the Interior that his name is Metatron.)

R. Yishmael says, How big is the size of the King of Kings, Bless be He that He is covered with all of Creation.

His feet are filling all the Earth, as it is said, "Heaven is my throne and the Earth my foot-stool".

For the readers information. A Myriad=10,000

A Parsang=4 miles

The sole of His foot is: three thousands Myriads of 1000 Parsangs – meaning[(3000 x 10,000) x 4 (a Parsang)] x 1000 = 120 billion miles each sole.

And from His sole to his ankle is; 1500x Myriads 10,000 x of 1000 x of Parsangs (4 miles) = 60 billion miles in height.

From His ankle to the knee: 19000 Myriads x of thousands parsang x 4 miles= 760,000,000,016 or 760 billions and 16 miles.

From His knee to hip: 12,000 Myriads of thousands and 10.04 Parsang = [12,000 x10,000x 1000] x (10.04)x(4) = 480,000,004,016 or 480 billion and 16 miles.

From His hip to His neck: 54,000 Myriads of thousands Parsangs [(54,000 x 10,000) x 1000] x 4=216,000,000,000.00 or 2trillion, 160 billion miles.

The hight of His neck: 13,800 Myriads of thousands Parsang, or [(13,800 x 10,000)x(1000)] x4 = 552,000,000,000 miles or 552 billion miles.

The diameter of His head is: 300,000 Myriad and 33 1/3 Parsang or[(.300,,000). x (10,000)]x[(33.3)x 4] = 399,600,000,000,000.00. what the mouth cannot pronounce and the ear cannot hear.

His beard: a Myriad of thousand five hundred Parsang or (1500x 10,000) x 4 = 60,000,000 miles .

All along each limb a secret name of THE CREATOR is mentioned. The look of His face, as the look and the Jade, like a spirit and in the form of a soul, that there is no creation, that can be mentioned. His body like a Topaz brilliance, light, very clean in the midst of darkness, clouds, and fog are surrounding Him. All ministering Angels, and Archangels are in front of Him.

We do not have a real measure, but only names that are known to us.

The pupil of the right eye: Myriad of thousand five hundred Parsang or (10,000 x 1500) x 4 = 60,000,000 miles or 60 million miles. the white

of His eyes: 22,002 Myriads of Par sang or (22,002 x 10,000)x=4880,080,000 milesor 880million,80,000miles.

From shoulder to shoulder: 16,000 Myriads of thousands Parsang, [(16,000 x 10,000) x(1000)] x 4 = 640,000,000,000 miles or 640 billion miles.

from His right arm to His left arm: 12,000 Myriads of thousands Parsang [(12,000 x 10,000) x 1,000] x4= 480,000,000,000 miles or 480 billion miles.

the fingers of His hands: 1500 Myriads of thousands Parsang or [(1500 x 10,000) x 1000] x 4 = 60,000,000,000 miles or 60 billion miles each finger.

The palm of His hands: 4000 Myriad of thousand Parsangs or [(4000 x 10,000) x 1000] x 4 = 160,000,000,000 miles or 160 billion miles.

His toes of the right foot: [(1000Myriad x 10,000) x 1000] x 4 = 40,000,000,000 miles or 40 billion miles, and it goes on and on. Rabbi Nathan, student of Rabbi Yishmael said that the total height of the Creator is RAVO, Myriad of Myriads of thousands Parsangs.

Translates this to (Myriad x 10,000 = RAVO){[(10,000 x 10,000)] x [(10,000) x (1,000)]} x 4 miles = 4,000,000,000,000,000 or in other words 4000 trillion miles or 6800 light years or0 680 0 ASTRONS is the height of the universe.*(The numbers are short due to in the ancient times,there were no such numbers available,but here every number has a very hidden secret of creation and or THE CREATOR blessed be HE.)

When Rabbi Nathan mentioned all the above measurements, Rabbi Yishmael, in the presence of Rabbi AKIVA, said: "Anyone who knows these definitions of our Creator, the praises of the Holy One He, and that He is covered with all creation, is guaranteed to be from the world to come, and will live longer in this present life.

I cannot mention the secret names of the Creator and they are numerous. All of these secrets are mentioned in the Song of Songs (4:13). His cheeks are like a bed of spices; like banks of fragrant flowers: His lips like lilies, dropping flowing myrrh. His hands are like rods of Gold set with emeralds: His belly is polished ivory overlaid with sapphires. His legs are pillars of marble, set upon sockets of fine gold: His countenance is like the levanon, excellent as the cedars. His mouth is most sweet: And He is altogether lovely. This is my beloved, and this my friend, O daughters of Yerushalaim.

All these names have very definite secrets that will reveal more than I am in a position to reveal. These numbers show the Creator's greatness. Help us understand that our galaxy is an atom in a cell of the Creator's entity, Blessed be He.

So we are so insignificant that we are like a sub-atomic virus. The good people are: The antibodies that are fighting the virus. The bad people are: THE VIRUS,that are those who cause our world wars, battles, crimes, etc., in the body.

IT is the reason that the good should,stick together to fight the virus so that we keep the body sane. If the virus wins, this atom is destroyed, and that is the cycle of the 6000 per life cycle of the earth.

Even to explain the media communications, would have influence on bodies,or antibodies. We live in 4 parallel worlds, EMANATION, CREATION,

FORMATION, and DEED. Each one is divided into 10 levels, and each of the 10, to10 more, infinitely.

There are 4 main components of the human body. The bone denotes Deed, the hard material that we so much believe and live by. The muscle, equals

Formation. The Blood vessels equals Creation. The skin equals Emanation. So, we live in the bone dimension or the lowest of them, and people are trying to understand the Creator, and find Him, make images of Him.

We are trying to reach other planets and galaxies, when we should be trying to investigate our own body, that has all the secrets of the universe. Billions of life forms are within us. Why look for extra-terrestrial life when we don't even know our own inhabitants. The only way to communicate with the Creator, blessed be He, is through spiritual contact. Spiritually, we can reach the Heights, that we will never be able to reach through the physical. Let's try to eliminate the virus, and try to cure it, together, so that the body will be pure. Remember, if we make it hard on the body, the body will make it hard on us.

To comprehend what I have mentioned above,let us say that you have a broken bone,you will feel the pain in you whole body,so let not live the bone,the hard core world that we live in decay.

To understand the illusion of our present life, let me give some examples of what we Are. We are like a person who is being hypnotized. The hypnotizer can make him believe that he is a rabbit and he will act accordingly. The same thing occurs with us. We are an antibody or virus so we are under an illusion of our actual form.

The Book of Yetzira (Formation) said that the world is not the place for the Creator, but the Creator is the place for the world and in many instances, G-d is called MAKOM (place). So, this corroborates with our theory.let's explain our life through this theory - So the good guys are the antibodies and the bad guys are the virus, and the world wars happen only when the virus becomes too big, so there is a loss of life or antibodies, and the virus is killed like Hitler, Adolf Hitler, in Hebrew has a value of 357 or the VIROSIS in Hebrew is 357 + 18=375. CHAY =18 is the breadth of life that anyone has as mentioned in the first chapter. Therefore, he was a Virosis, meaning any disease caused by a virus, so he was a virus, so our particle (Earth), or our Atom Galaxy, has a contamination of a virus, and it was contained.

Prisons = Are, contention of virus or in a bad country, a contention and illumination antibody.

Women having a baby = A potential antibody of virus depending on directions of the parents.

Media = Influence to one of the two.

President = Head of the virus or antibody.

Planes, cars, transportation = Ways of spreading around the particle.

Animals, plants = Nourishment for antibodies or virus.

And all of us are some kind of virus or antibody. We are under the illusion of a life, task or situation.

Police = Controlling agents against infestations.

And if we control the virus and stay virus free, and the only way of communication, toward the Creator, is by prayers, spiritual approach, because physically it's impossible. So when we are converted in Energy Intelligence, we become one with the Creator, and the virus is destroyed.

Albert Einstein developed his general theory of relativity. He found and predicted that the universe is expanding. The body expands too. This is not blasphemy but I am quoting the clues and the vision of a prophet, as I mentioned in the beginning of this chapter

THE REAL AGE OF THE EARTH

הגיל האמיתי של כדור הארץ

CHAPTER II

The Real Age of the Earth
הגיל האמיתי של כדור הארץ

We are going to cite Bible quotes. In Genesis, we find the ages and life span of Adam and his descendants. We know for a fact that Adam existed, because we have in Israel, his tomb. He is buried with our fathers, Abraham, Isaac, and Jacob in the tomb of Machpela (in Hebrew). We therefore, know for a fact, he existed. But the Bible brings us information of life span of the following:

Understanding that 1 day of the Creator is 1000 years. This book of the chronicles of Adam or in this case, Earth. (in Hebrew "Adam" is "Earth").

(Years He Lived x Days) x Years

10 Adam (93Ox365)x 1000= 339,450,000 years

2) Seth (912x365)x l000 = 332,880,000 years

3) Enosh (910x365)x 1000= 330,325,000 years

4) Kenan (910x365)x l000= 332,215,000 years

5) Mehalel (895x365)x l000 = 326,675,000 years

6) Yered (962x365)x l000 = 351,130,000 years

7) Enoch (365x365)x l000 = 133,225,000 years

8) Metoushelah (965x365)x l000= 353,685,000year

9)Lemech (777x365)x 1000 = 283,605,000 years

10) Noah (950x365)x 1000 = 346,750,000 years

The Flood.* TOTAL3,129,940,000 years

* In 6000 year cycles.

All those 10 different names indicate 10 periods that the Earth was inhabited and in between desolation. I'll mention it in the next page. Those were the 3 billion, one hundred twenty-nine million nine hundred forty thousand years of inhabitants in our Earth up to the flood era..

Now we will take the second clue in the age before each man of the 10 fathered. Fathered meaning that before they had a child. The earth desolate, no life form on it.

1) Between Adam & Seth

(130 x 365) x 1000 =47,450,000 years

2) Between Seth & Enosh

(105 x 365) x 1000 =38,325,000 years

3) Between Enosh & Kenan

(90 x 365) x 1000 =32,850,000 years

4) Between Kenan & Mehalael

(70 x 365) x 1000 = 25,550,000 years

5) Between Mahalael & Yered

(65 x 365) x 1000 =23,725,000 years

6) Between Yered & Chanoch

(162 x 365) x 1000 =59,130,000 years

7) Between Chanoch & Metouselah

(65 x 365) x 1000= 23,725,000 years

8) Between Meltauselah & Lemech

(187 x 365) x 1000 = 68,255,000 years

9) Between Lemech & Noah

(182 x 365) x 1000 =66,430,000 years

10) Between Noah & Shem

(500 x 365) x 1000 =182,500,000 years

TOTAL years of desolation 576,940,000 years

Genesis 6:23) He saw that man's wickedness on Earth was increasing. Every impulse of His innermost thought was for evil all day long. G-d regretted that He has made man on Earth, and He pained to His very core. G-d said, "I will obliterate humanity that I have created from the face of the Earth, Man, livestock, land, animals and birds of the sky. I will regret that I created them." But Noah found favor in G-d's eyes.

Here we see how G-d got angry and destroyed the Earth, but not one time but 10 times. So, there were inhabitants for 3,129,940,000 years, there was life on Earth, and 567,940,000 years, desolation, or 3 billion, 697 million, 880 thousand years, is the beginning of creation of our Earth.

After Noah, G-d did not destroy the Earth any further. And that is the explanation of this 10 names and life span, to show us the time of creation. By the same token, we are in our life cycle in the year 5752 of this new creation. The new Adam, the new world and not Adam Kadmon (prehistoric man). There are so many depths to the Bible that all secrets of life are hidden on it's very depths. Some people just decide that they don't want to know. The secrets are there for you to reach, not just to leave it there. When you buy jewels, you wear them. You do not keep them in a safe only.

EXTRA TERRESTRIALS
UFO'S
חייזרים ועצמים
בלתי מזוהים

CHAPTER III

EXTRA TERRESTRIALS, UFO'S

Who are they? Do they exist? Are they a product of our imagination? Are they a military secret weapon? I will proceed to present three theories that are similar in one aspect. They are all theories due to the fact that there is no definite proof. This phenomenon has been seen through the ages, so this discards one possibility, "the military."

Theory One

Extraterrestrials are our past. Our forefathers from before this life cycle, at least six thousand years ago. There was a life more sophisticated than ours in technologies and awareness.

People were like us. Humans, not from another planet. " Kohelet", King Solomon said, "There is nothing new under the sun". We could understand from it, that all life that we are living today has been once before. We have various evidence showing that theories, like landing fields dated 10,000 years ago, a skull make of man a man-made fiber, perhaps of an android, the pyramids and it's secrets and so on.

There was a nuclear apocalypse, a world war, some several thousands years ago. A minority survived this nuclear catastrophe. They have a natural hiding place, buffered and protected from radioactivity, "The Bermuda Triangle". In underwater caves, or the City of Atlantis,they have generators, air purifiers, water desalinization systems, and all life under the ocean floor.

They did survive the nuclear war, but because some might have been radioactively contaminated. New generations were born deformed. Close encounters of the third kind, described those creatures as a general human form, but with features, as bigger heads, longer arms, some shorter.

These deformations are due to the radioactivity as seen in Hiroshima, new born babies with all kinds of deformities. In this case those "Aliens" are not speaking, they communicate through telepathy. Maybe due to the same radioactivity, they lost speech and hair, but developed telepathy. Also, for not being under the sun for generations, their paleness, no hair, etc...

We read and heard so many stories about the Bermuda triangle, boats disappearing, airplanes, or compasses go wild, when they are in the Bermuda Triangle area, those force fields or energy fields are generated by those "Aliens" to repulse explorers and curious people from finding them.

Finding them, would signify, finding technology. Political struggle for power and again wars to control the Earth and the eminent destruction of it. The abduction of some of "our humans" are to follow-up the progression of the human, mentally and physically. They had a longer life span than ours, being 70 years average, theirs being probably hundreds of years as mentioned in figure 9.

Genealogy, like Adam. He lived nine hundred and thirty years. A life span was reduced after the "great flood" to two hundred and following down to one hundred sixty-five, etc., till us, 70 years average. Animals were bigger, giants, as in the Jurassic era, and radiation created a change in the D.N.A. and all those giants disappeared and changed shape. . The new species were reduced in size. The only one that kept a big size was the elephant, in the area of Africa, probably radiation did not reach all areas of the globe. So the great flood could have been a nuclear holocaust, but again, not eliminating the actual great flood in the Bible, because this one was in our life cycle of the last 6000 years, because of all of the clues of the Bible speaks for all times, past, present and future.

"There is nothing new under the Sun". The presence of aliens, appearing specially above military installations to follow and try to control a possible new nuclear holocaust. The reason they disappear from the radar and show up from nowhere is due to the fact that they come up from underwater in the Bermuda Triangle and when they disappear, they plunge back into the water. Today, the human has entered the ""Aquarius" era, the Kabbalah mentions that this is the era, where all technology and the secrets of all life will be revealed to humans and the advancement will be great.

Today, probably, there are "Aliens" between us, reconstructed genetically to our likeness. They are trying to cause peace on earth and by the same token to get jobs in key positions close to nuclear places. After each close encounter, people in black, showed up and talked to those people who had that experience, to see what their reaction is, are they ready for a final encounter?

The Second Theory

The same as the number one, buy they are coming from our future. A nuclear war will happen, and very few will survive, and we will become as described in the above theory, due to radioactivity. We are coming back in time to the past, starting from the beginning of ages to try to analyze and find where things started to go wrong, and try to undo it, but it is very delicate, it could cause a "Paradox", and that will be a catastrophy, and will create a chain reaction of changes, and all future could be in jeopardy, resulting in unlimited changes, that will maybe accelerate the destruction of life with no possible way back, as they are trying to investigate.

The Third Theory

Real extraterrestrials, from other planets, that are trying to contact us. But who are they? How can they travel from places, millions of light years away? Are there other forms of life on other planets? The answer is that it is possible.

How? Well we know that all molecular structure is an agglomeration of atoms. Atoms are composed of protons, neutrons, and electrons and many more particles. The gluons that are keeping them together, is the intelligence energy, as explained in past chapters. We could dematerialize or materialize to our likeness, if we could control our inner and surrounding intelligence energy, that keep together all forms of life.

When traveling incredible distances, the Kabbalah explains that time and speed is a human definition, there is no such thing as speed but only in our minds. The way these U.F.O.'s travel is simply by controlling the inner and surrounding intelligence energies(electrons) and ordering them to leave their atoms.

Leaving blank air atoms in place and their intelligence energy transferred instantly to any distance in the universe. Once in the desired place, the energy intelligence will rematerialize in local blanks atoms in the air space, but there are no vacuum spaces of atoms, but larger atoms or smaller atoms. A "black hole" could be a complete different surrounding of various layers of different types and sizes of atoms, and each size of atoms could signify an era of time in space. It could signify the dark tunnel perceived in the N.D .E. (near death experience) by thousands of people, and they go through their past life and then to the future, in spiritual form. So, this could explain the theories of the black hole, for travel in space and time, past or future. This way those aliens come to visit and investigate our planet, our life. Some might have incorporated our cities and live a life here on earth. Our galaxy is only a grain of sand in the universe, as mentioned in chapter 7, only an atom of the body of our Creator.

What should we do about those extraterrestrials? In my opinion, contact them in any way possible, they are a lot more sophisticated and intelligent than us. We are as compared in chapter 7, only the bone of the body, the center, the core, the hard part of the body, we are materialistic.

They probably come from the muscle, blood vessels, or the skin atoms of the body of the universe. If a bone is decaying it affects all other layers of the body causing pain, and the reason the "Aliens" come to our planet (bone) to correct or cure our planet and find and correct our deficiencies, so the body will be healthy.

These are the three theories about U.F.O.' s and alien beings, but the best advice is to worry about our planet "the Bone", to keep it in good shape, healthy, so the whole body will stay healthy, and the aliens want to contact us. Let's be ready to hear and learn from them. They know better, they will never invade us, they could not live in our environment. Let's keep our minds open.

Other Clues From the BIBLE

CHAPTER IV

OTHER CLUES FROM THE BIBLE

Some people will argue that creation was with Adam. I do not argue that, but that was the last creation and not the first. In Genesis, it is mentioned twice the creation of man, first (1 :26), G-d said, "Let us make man with our image and likeness". (1 :27), G-d created man with His image. In the image of G-d, He created man with His image. In the image of G-d, He created him, male and female, He created them. G-d blessed them. Here

G-d does not talk about Adam Harishon (Adam first man) but (Adam Ha Kadmon),3.6 billion years ago. And it was the sixth day of that creation.

(Genesis 2:4) "There are the chronicles of heaven and Earth, when they were created, on the day G-d completed earth and heaven", here heaven and Earth are repeated twice, one for Adam Kadmon, prehistoric man, one in our life cycle of Adam Rishon (first man).

(Genesis 2:5) "There was because G-d had not brought rain on Earth, and there was no man to work the ground". Here it-shows that the Bible does not refer to Adam Harishon because He would be in Gan Eden* and did not have to work the ground if not for his sins. The Bible talks about creation first in the 6 first days and rested on the 7th. But of the first creation, and not this last one.

(Genesis 2:7) "G-d formed man of dust of the ground, and breathed life into his nostrils, a breath of life. Man became a living creature". Here the Bible talks about Adam H.arishon (Adam the first) and this is when our calendar starts ticking - year 1.

It will restart after each 6000 year circle and one thousand years desolation or the sabbath of creation. (Genesis 2:10) "A river flowed out of Eden to water the Garden, from there it devided and became four major rivers.

The name of the first river is, Pishon it surrounds the Gan Eden meaning Heaven (Garden of Eden). entire land of Havilah.

The name of the second river, is Gihon it surrounds the land of Cush.

The name of the third river is, Chidekel which flows to the east of Ashur.

And the 4th river is Prat. This one goes nowhere.

Here again many are the clues.

One: Textually.

Second: As I mentioned in the beginning of the book.

Third: Here, since we are talking of the universe being in a form of man, Eden is the Essence of G-d, the Ein-Sof, the light, the brilliance of the King of Kings, then the;

1st River Pishon surrounding the entire land of Havilah, would be the skin surrounding all the body.(according to my theory).

2nd River Gihon, it surrounds the land of Cush, meaning the arteries,veines that surrounds the muscles.

3rd River Chidekel,which flows to the East of Ashur,thus the muscles.

4th river Perat, it does not mention anything about it, because it is the core of the bone and it is surrounded by all the above and also they form the 4 dimensions, EMANATION, CREATION, FORMATION, and DEED.

For a reason our sages said that there are 70 facets of the Bible and all truth, that's the beauty of it, every facet is a secret by itself, showing a new meaning or level of understanding for the whole bible.

(Genesis 3:9) G-d called to the man, and He said "Where are you trying to hide?" (Genesis 3:10) "I heard your voice in the Garden," replied the man, "and I was afraid because I was naked, so I hid." (Genesis 3:11) G-d asked, "Who told you that you were naked? Did you eat from the tree which I commanded you not to eat?"

Here we see clearly that he had to eat from the tree of knowledge to find that he was naked, he had to be told like G-d asked him. Meaning that we are clothed with illusion but we do not see the true reality.

We live in a world of illusion, where everything has a material cover or shell, a task, and all is an illusion, as I explained in the beginning of this chapter, and if we study the Kabbalah (the tree of knowledge), we will find our way back to Eden, eternal life, to incorporate G-d' s unity.

In Genesis, it mentions one time the tree of knowledge, (Genesis 2:9) to Adam alone, there is the knowledge that man only should know due to the terrible secrets, and the woman was not created yet. Then after being punished and banished from Eden (Genesis 3:22) "He drove away the man, and stationed the Cherubim at the east of Eden along with the revolving sword blade, to guard the ["Path of the tree of life"].

Here He mentions the tree of life, and not the tree of knowledge, here is the "Path of Life" our predestination and the revolving blade, as I mentioned before, are the Seven Planets, Jupiter, Saturn, Mercury, Venus, Mars, Moon and Sun and zodiacal star formation that influence our Path of life, and make us forget our past knowledge, and Cherubims, to guard, are: One is the moon, and second is the planet Venus. (1) (See figure 22).

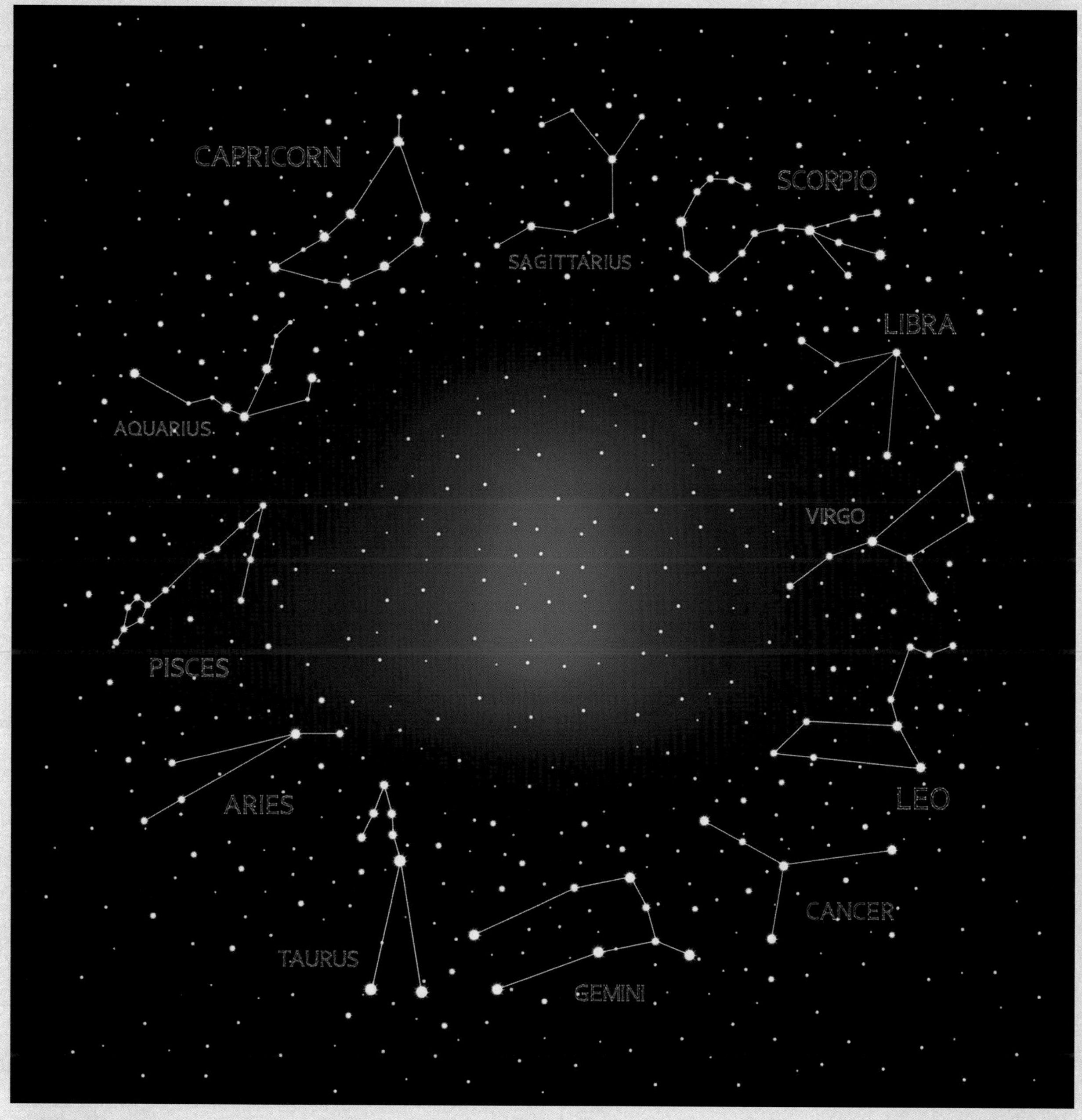

FIGURE No .22: One of the CHERUBIMS forming part of the Revolving BLADE.

The other CHERUB is the SEVEN PLANETS,the other revolving blade. These are the reason we forget our past lives and affect our present one,on our behavior and DESTINY.

(Genesis 4:) 1) "The man knew his wife Eve (CHAVA). She conceived and gave birth to Cain. "Here it mentions the man knew his wife (feminine side), Moon (Yareach), in Hebrew, Cain is written (Kouf-K) (Yud-Y) (Noon- N). The Kuf stands for Klipah, meaning shell, or the shell of negativity conceived from Yareah, from (Yud) meaning the Moon's influence, and (Noon-N) for Noga (Venus), so the shell of negativity is formed from the Moon and Venus influence and those are the 2 Cherubims.

(Genesis 4:2) "She gave birth again, this time to his brother Abel". In Hebrew it is Havel and Havel in Hebrew means, vanity. If we analyze both together, the man knew that his negative side or shell is caused because of the influence of the stars, Venus, and the Moon, and both causes vanity, and vanity is one of the four negative qualities that man has to avoid.

"Abel became a shepherd, while Cain was a worker of the soil." (Genesis 4:3) Abel or Vanity (Roeeh Tson) became a shepherd, in this case it means that vanity Roeeh, it also read Raah, meaning evil wickedness or nourished wickedness. Tson (Flock), TS for (Tsedek or Jupiter) Alef, for (Elil-god) (N-Noon) for (Nevel degeneracy) in other words, people of that time degenerated in wickedness worshipping a false god, the planet Jupiter, and Cain, the (shell of negativity) was a worker of the soil, meaning also in Hebrew, worship of idols, and G-d destroyed those generations and the Era ended.

The last era before Noah. An Era (Genesis 4:3) meaning the Era of Adam Kadrnon. Now the Bible goes back to the actual Cain and Abel that are really, one a shepherd and the other a worker of the soil, and explains of Abel and Cain sons of Adam Harishon or the first of this cycle of life on Earth, year One of this calendar being today 5752.

Here we are expanding on a 2nd explanation as Abel being the good side, Abel is Havel in Hebrew, if you change the order of the letters, it changes to Halev (the Heart) meaning when the thought reflects the Heart, and Cain being the evil side, here Cain in Hebrew, Kayin (K-Kouf) Klipah (Shell) (Y -yud) for Yetzer (evil incarnation) (N-Noun) Nachash (serpent), also Nahash = 348, satan = 349, after he pushed Eve to sin, his was added the one, number one evil.

One gives the best and the other says, why the best, I'll give what I want, like an internal argument of one lesson. When the Bible says (Genesis 4:3) Cain brought some of his crops as an offering to G-d. Abel also offered some of his first born of his flock from the fattest one." Abel sought and was for G-d without reserves. (Genesis 4:4)

"G-d paid heed to Abel and his offering, but to Cain and his offering, He paid no heed. Cain became very furious and depressed. G-d said to Cain, why are you so furious? Why are you so depressed? If you do good, will then not be special privileges? And if you do not do good, sin is crouching at the door, it lusts after you, but you can dominate it."

Here, the Bible shows the good winning over the evil,meaning the good side overcame. "G-d to Cain and his offering, He paid no heed" and here to the evil thought did not succeed. We understand it specially when He says that "Sin is crouching at the door, it lusts after you, but you can dominate it," it means textually that evil is always present within us but we can dominate it".

(Genesis 4:8) "Cain said to his brother Abel, then, when they happened to be in the field, Cain rose up against his brother Abel, and killed him, G-d asked Cain, "Where is your brother Abel?" Here's what will happen if evil overcomes good. When you kill the good in you, the following will be your punishment. (Genesis 4: 13) Cain said, "My sin is too great to bear! Behold, today you have banished me from the face of the earth, and I aim to be hidden from your face. I am to be restless and isolated and isolated in the world, and whoever finds me, kill me."

Here the punishment for evil is being banished from the face of the earth and the "Earth swallowed that is to go to Gehinom -גהינום(Purgatory), isolated because in Purgatory one is alone, and where he says to hide from your face the only way to hide from G-d, where he is in Purgatory that the Holiness of G-d would not enter "Sheol-שאול", the underworld.

(Genesis 4: 15) "Cain went out from before G-d's presence in a world of darkness. He settled in the land of Nod" the only place he can be, not in the place of G-d is the underworld. The name Nod is composed from the (Noon-נ - N) for (Nozli-נוזלי) Liquid, then O in Hebrew is Vav-ו and in this case means (and) and Dalet-ד for domem-דומם (inanimate) and that is what I explained before, becoming fuel, going down toward the core of the earth, and it is one of the worst deaths, where there is no return, and that is the punishment also for a person who kills someone else, he doesn't kill only him, but generations to come. As Cain killed half the world.

ADVISES FOR A BETTER LIFE FROM "ETHICS OF THE FATHERS"

CHAPTER V

ADVISES FOR A BETTER LIFE FROM

"ETHICS OF THE FATHERS"

"Let your house be a meeting place for sages." (Stick with good people and you will learn only good things).

"Provide yourself with a teacher; acquire for yourself a friend; and judge every person favorably. (A teacher is anyone whom you can learn from. Acquire a friend for yourself; look for a good friend, with good feelings and never judge a person, because you will be judged by others in return).

"Keep away from a bad neighbor; do not fraternize with a wicked man; and do not abandon belief in Divine Retribution." (If you fraternize with bad people, you will share their wicked destiny; whatever happens to him, you will share it, don't abandon belief in Retribution, because behind every bad experience, good will result).

"When sitting as a judge; do not act as a counselor; when the litigants stand before you, regard them both as guilty; but when they leave, having accepted the judgment regard them both as guiltless." (In other words, don't form an opinion before you have heard both sides and testimonies, and do not take as a witness a pervert or those involved in the same case).

"Examine the witnesses thoroughly; and be cautious with your words, lead through them the witnesses or the litigants learn to speak falsehood."

"Love, peace and pursue peace, love your fellow creatures, and bring them near studies of Bible." Because everything there is, is learned from the Bible and if you do good, good will you receive, (Mida Keneged Mida-מידה כנגד מידה) (Measure against measure).

"If I am not for myself, who is for Me? And if I am only for myself, what am I? And if not now, when?" (Do not wait for others to do your deeds,

because if you don't help yourself, you are a selfish person who thinks only of himself, so he will receive measure against measure, and if you do not start now, then when? Because there is no better time than now, because if you hesitate, it will never be done).

"Set a fixed time for study", it will be feasible to study, because as I said, every action begins with a thought, so, if in your mind it is set, the time will be dedicated to study. When you say little and do much people will appreciate you more. Because when you say I will do this and that, and then you don't do it, people will see that you are not a man of conviction, and reputation sticks, one way or another.

Always have a smile when you receive a person or a guest, because your cheerful face, will make the person feel good, and his feeling will make you feel good in your inner self. The contrary will make that person offended and will feel bad, and that will reflect in your own Intelligence Energy.

"Provide yourself with a teacher and free yourself of doubt; and do not do guesswork."(When you have doubts in general, find someone who really knows about that matter, because guesswork will lead to a chain of mistakes that will make you look bad, even though your intentions were honorable).

"All my days I grew up among the Sages and did not find anything better for one's person than silence; not study but practice the essential thing; and whoever engages in excessive talk brings on sin." (Silence is of gold, because whatever you say will come ..back against you only if it is good, it will come back to your advantage.

If the talk you intent is futile, it is better to stay silent. You will not gain anything but increase the possibility of a future problem. To study and not to practice, is worse than not knowing and not doing).

"The world endures by virtue of three things - Justice, Truth and Peace, as it is stated: Administer truth and the judgment of peace in your gates." (If the whole world respected those principals, there will not be wars, crime, and all humanity will be a brotherhood, and the Messiah will come).

"Which is the right path that a man should choose for himself? That which is honorable to himself and brings him honor from man?" (In other words, choose a path of good deed, toward helpless children, orphans, charity, old people, sick, and be a scholar that will bring you all the honors a man can wish for, and a fantastic feeling of satisfaction, that no money can buy).

"Reflect upon three things and you will not come to sin: Know what is above you and Eye that sees, and Ear that hears, and all your deeds are recorded in the Book." (As I explained in another chapter, the Creator is everywhere, and there is no place to hide).

"Be wary of those in power, for they befriend a person only for their own benefit; they seem to be friends when it is to their advantage, but they do not stand by a man, in his hour of need."

"Fulfill His will as you would your own will, so that He may fulfill your will as though it were His will, so that He may set aside the will of others before your will." (If you do G-d's will even though you do not comprehend it, it is for your good, because G-d does not request from us anything that at the end, will not be for our good).

"Do not separate yourself from the community; do not be sure of yourself until the day you die; do not condemn your fellow man until you have stood in his place, do not make an ambiguous statement which is not readily understood in the belief that it will ultimately be understood; and do not say, "When I will have free time I will study," for perhaps you will never have free time.

"He saw a skull floating on the water; he said to it: Because you drowned others, they drowned you; and ultimately those who drowned you, will themselves be drowned." (And this is "Mida Keneged Mida" or measure against measure).

We have a good example at the end of the Passover prayer, after dinner, we say a story for the children: "The goat that my father bought for two zuze (coin of that time): The cat came and ate the goat that my father bought for 2 zuze, then the dog came and bit the cat that ate the goat.. The the stick came and hit the dog that bit the cat that ate the goat. Then equally the fire burned the stick that hit the dog,that bit the cat that ate the goat. Then, the water extinguishes the fire that burned the stick etc.., then the cow drinks the water that extinguished the fire ect.., then the shokhet(the slautherer) kills the cow that drank the water etc..., then the angel of death kills the shocket(slautherer) that killed the cow etc..., than G-d Blessed he He, kills the Angel of Death,that killed the slautherer,that killed the cow,that drank the water,that extinguished the fire that burned the stick that hit the dog that bit the cat that ate the goat that my father bought for two zuze..

All the story is really a parable that G-d sends always someone to punish someone else, because that man that punishes is himself going to be punished and so on, to the ultimate; G-d's punishment.

This creates a vicious circle. In order not to fall into it, do not take the law, or revenge to your hands. also, let THE ULTIMATE JUDGE to do justice. So you won't fall in to the vicious cycle.

"Increasing flesh, increases worms (in the grave); increasing possessions, increases worry; increasing Torah, increases life, increasing assiduous study, increases wisdom, increasing counsel, increases understanding, increasing charity, increases peace. One who has acquired for himself Torah knowledge, has acquired for himself life, in the world to come. (Here in other worlds, material or physical is only to the detriment to oneself but spiritual only increases your Intelligence Energy and that is the only thing that you will take with you when you depart from this physical world, the world of illusion).

"Let the honor of your fellow, be as dear to you as your own and do not be easily angered. Repent one day before your death. Warm yourself by the fire of the sages, but beware of their glowing embers lest you be burnt - for their bite is the bite of a fox, their sting the sting of a scorpion, their hiss is the hiss of a serpent, and all their words are like fiery coals." (Here is, Love thy neighbor as thyself.

To repent even one day before death, in order at least to return and have a second chance, as a human being and not a lower form of Intelligence, of course, a person one day before dying, is sincere, we assume. Come close to sages to learn from them, but if you are trying to do it for evil purposes, you will get harmed because those people are in a "higher plane" than your own and benefit from higher protection.

"The evil eye, the evil inclination, and hatred of one's fellow, drive a man from the world." (It does drive man to the underworld, the world of darkness)

"The day is short, the work is much, the workmen are lazy, the reward is great, and the Master is pressuring." ("The day" is our life span. The work, is the study of the Kabbalah, and Torah, it is very long, and we are lazy, we do not dedicate enough time to study, occupied with material need, the reward is the biggest you can think of, it is life, or eternal life, to incorporate the "Ein-Sof' the Endless Light, the Master is G-d our Creator).

"Reflect upon these three things and you will not come to sin: Know from where you come, and to where you are going, and before whom you are destined to give accounting." "From where you come" from a putrid drop, (of semen); and to "where you are going" - to a place of dust, maggots and worms; (the tomb when a person dies); and before whom you are destined to give accounting - before the supreme King of Kings, the Holy One, blessed be He. (If we have those three questions during our life time and their answers, we will be aware that we are constantly connected to the Creator and there is no way to hide, and everything you do positive or negative in the balance the "day you will leave this physical world, as I explained in previous chapters).

"Pray for the welfare of the government, for were it not for the fear of it, men would swallow one another alive."

"Anyone whose fear of sin comes before his wisdom, his wisdom will endure; but anyone whose wisdom comes before his fear of sin, his wisdom will not endure."

"And anyone whose good deeds exceed his wisdom, his wisdom will endure; but anyone whose wisdom exceeds his good deeds, his wisdom will not endure."

"The sleep of the late morning, wine at midday, and sitting in the gathering places of the ignorant, drive a man from this world." "Laughter and frivolity accustom a man to lewdness. Tradition is a fence around the Torah (Bible); tithes are a fence for riches vows are a fence for abstinence; a fence for wisdom is silence."

"Everything is foreseen, yet freedom of choice is granted; the world is judged with goodness, and everything is according to the preponderance of good deeds." (We are predestined, but following the Commandments will change any fatal predestinations-see previous chapter).

"Everything is given on collateral and a net is spread over all the living; the shop is open, the shopkeeper extends credit, the ledger is open, the hand writes and whoever wishes to borrow, let him come and borrow; the collectors make their rounds regularly, each day, and exact payment from man with or without his knowledge (of his debt), and they have on what to rely; the judgment is a judgment of truth; and everything is prepared for the feast." "The shop is open" (is our possibility to choose good or evil,("the hand writes" (your credit or debt are accounted for) "and whoever wishes to borrow" (Evil or bad deeds, because good is credit and not debt); the collectors make their rounds regularly (measure against measure). "The judgment is a judgment of truth," (you can lie to yourself but not to the Creator, because He knows your most inner thoughts), "and everything is prepared for the feast" is when you die, you go in front of the Creator depending on what you accumulated during your life time. If good, you will go to higher levels. If bad, the underworld. See previous chapters.

"Anyone whose wisdom exceeds his good deeds, to what can he be compared?" To a tree whose branches are numerous, but whose roots are few, and the wind comes and uproots it and turns it upside down; as it is stated: And he shall be like a lonely tree in arid land and shall not see when good comes; he shall dwell on parched soil in the wilderness, on salt-land, not inhabitable. (Jeremiah 17:6) But anyone whose good deeds exceed his wisdom, to what can he be compared? To a tree whose branches are few, but whose roots are numerous, so that even if all the winds in the world were to come and blow against it, they could not move it from it's place; as it is stated: And he shall be like a tree planted by waters, toward a stream spreading it's roots, and it shall not feel when the heat comes, and its foliage shall be verdant, in the year of the drought it shall not worry, nor shall it cease from yielding fruit. (Jeremiah 17:8("The laws pertaining to bird - sacrifices and calculation of the onset menstruation these are essentials of the Torah Law; (see previous chapters on Kosher and causes of after menstruation pregnancy).

"The calculation of cycles (astronomy) and numerical computations of Hebrew words (see our chapter on the power of the letters) are Condiments to wisdom."

"Who is wise? He who learns from every person, as it is stated: From all those who have taught me I have gained wisdom." (We even learn from a fool, what not to do indeed, your testimonies are my conversation. (psalms 119:9(."

"Who is strong? He who subdues his evil inclination, as it is stated: He who is slow to anger is better than the strong man, and he who masters his passions, is better than one who conquers a city (proverbs 16:32)."

"Who is rich? He who is happy with his lot as it is said: When you eat of the labor of your hands, happy are you and it shall be with you" (psalms 128:2)."

"Happy are you" - in this world, "and it shall be well with you" - in the world to come. (It is difficult to be happy with what we have, we always want more, than we will never be rich enough but he who is happy with his lot is the real rich, he doesn't suffer the disease of greed.)"

"Who is honored? He who honors others, as it is stated: Indeed, those who honor Me, I will honor, and those who despise Me, shall be degraded." (Samuel 12:30).

"Run to perform even an easy good deed, and flee from transgression; for one good deed brings another, and one transgression brings about another" (a good deed gives you satisfaction, and you want to feel that feeling again, but a transgression brings another one. As lying, to cover his misdeed and that will lead to others. (As killing someone, suddenly you found out that there is a witness, as you have to kill him too, and he told someone else, you have to kill him too, to no end.

"For the reward of a good deed is a good deed, and the recompense of a transgression is a transgression."

"Do not regard anyone with contempt, and do not reject anything, for there is no man who does not have his hour and nothing which does not have its place."

"Be of an exceedingly humble spirit, for the expectation of mortal man is but worms. (supra, Avot 3:1) (here is one of the four qualities that we should avoid and this one is pride).

"Whoever desecrates the Heavenly name in secret, punishment will be meted out to him in public; unwittingly or intentionally, it is the same in regard to the desecration of the name." (as mentioned in previous chapter the severity and power of the name of G-d).

"(A judge) who refrains from handling down legal judgments (but instead seeks compromise between the litigants) removes from himself enmity, theft, and (the responsibility for) an unnecessary oath, but one who aggrandizes himself by (eagerly issuing legal decisions is a fool, wicked and arrogant).

"Whoever fulfills the Torah in poverty will ultimately fulfill it in wealth; but whoever neglects the Torah in wealth will ultimately neglect it in poverty".

"Let the honor of your student be as dear to you as your own, the honor of your colleague as the reverence for your teacher as the fear of Heaven".

"Be the first to extend greetings to anyone you meet; and rather be a tail to lions than a head to foxes."

"This world is like an ante-chamber before the WORLD- TO-COME; prepare yourself in the ante-chamber so that you may enter the banquet hall."

"Do not placate your fellow in the moment of his anger; do not comfort him while his dead lies before him; do not question him of his vow at the moment he makes it; and do not seek to see him at the time of his degradation."

"When your enemy falls do not rejoice, and when he stumbles let your heart not be glad, lest the Lord see and it will be displeasing to him, and He will divert His wrath from him to you."

H"e who studies Torah as a child, to what he be compared? To ink written on fresh paper, and who studies Torah as an old man, to what can he be compared? To ink written on paper that has been erased."

"He who learns Torah from the young, to what can he be compared? To one who eats unripe grapes or drinks wine from his vat; while he who learns Torah from the old, to what can he be compared? To one who eats ripe grapes or drinks aged wine."

"Do not look at the vessel, but rather at what it contains; there may be a new vessel filled with aged wine, or an old vessel in which there is not even new (wine)," This is to learn that the appearance of a person, is not the important but his soul (wine) and the vessel is the body.

Those who are born are destined to die; those who are dead are destined to be judged, (Therefore, let man) know, make know, and become aware that He is G-d, He is the Fashioner, He is the Creator, He is the Discerner, He is the Judge, He is the Witness, He is the Plaintiff, He will thereafter sit in judgment, Blessed is He before whom there is no inequity, nor forgetting, nor partiality, nor bribe taking;(How can we bribe G-d? well we do good deeds, after a bad one as saying I stole money but I will share it with poor people.) and know that the grave will be a place of refuge for you for against your will you were created, against your will you were born; against your will you live; against your will you die; and against your will you are destined to give an account before the Supreme King of Kings, the Holy One, Blessed be He",

"Seven things characterize a stupid person, and seven a wise one, A wise man does not speak before one who is greater than he is, in wisdom or in years; he does not interrupt the words of his fellow; he does not rush to answer; he asks what is relevant to the subject matter and replies to the point; he speaks of first things first, and of last things last; concerning that which he has not heard he says, "I have not heard"; and he acknowledges the truth, And the reverse of these characterize a stupid person."

"There are four (character) types among men: He who says, 'what is mine is yours, and what is yours is mine', is an ignoramus; "what is mine is mine, and what is yours is yours" - this is the median characteristic:

"What is mine is yours. and what is yours is yours," is a pious, benevolent person; He who says, "What is yours is mine. and what is mine is mine", is a wicked person.

There are four types of temperaments: "Easily angered and easily pacified" – his loss is outweighed by his merit; "Hard to anger and hard to pacify" - his merit is outweighed by his loss; "Hard to anger and easy to pacify" is pious. "Easily angered and hard to pacify" is a wicked person.

There are four types of students

"Quick to grasp and quick to forget." - his advantage is canceled by his disadvantage.

"Hard to grasp and hard to forget." - his disadvantage is canceled by his advantage.

"Quick to grasp and hard to forget." - This is a good portion.

"Hard to grasp and quick to forget." - This is a bad portion.

There are four types among those who give to charity:

"One who wishes to give but that others should not" - he begrudges others.

"That others should give and he should not" - he begrudges himself.

"That he should give and others should too" - is a pious person.

"That he should not give nor should others" - is a wicked person.

"Whoever causes the many to have merit, no sin shall come through him"; "but one who causes the many to sin shall not be granted the opportunity to repent." Moses was himself the meritorious and the many to attain merit, therefore the merit of the many are attributed to him, as it is stated: "He (Moses) performed the righteousness of the Lord, and His ordinances together with Israel." Yarevam Ben Nevat himself sinned and caused the many to sin, (therefore) the sin of them many are attributed to him, as it is stated: "For the sins of Yarevam which he sinned and caused Israel to sin".

"Be bold as a leopard, light as an eagle, swift as a deer, and strong as a lion, to carry out the will of our Father in Heaven. (We have to be all that in order to know, what there is to be known and understand the Creator).

"Beauty, strength, wealth, honor, wisdom, old age, ripe old age, and children, are befitting the righteous and befitting the world as it is stated: Ripe old age is a crown of glory, it is to be found in the path of righteousness; (Ibid 16:31) and it says: The glory of young men is their strength, and the beauty of old men is ripe old age; (Ibid 20:29);and it also says: Grandchildren are the crown of the aged, and the glory of children are their fathers".

All those sayings are in away, constructing the fortification of your life path and they include practically, almost all the Commandments, their value is of diamonds, their words of gold; if you value your life as such. If you follow those advises, nothing bad will befall you, toward the Endless, the eternal life. "Eden"

THE WORLD OF MIRACLES

עולם הניסים

CHAPTER VI

THE WORLD OF MIRACLES

עולם הניסים

What is a miracle?.A miracle is something that happens and there is nothing we can logically or scientifically explain. Most of the miracles happen when there is a strong faith,in G-D,Saints being intermidiaries

To the Creator on our behalf.

But how does it work? Well of course the CREATOR is always at the end of each miracle. Why? Again it is because all creation exists because of him. We are part of His Essence,THE ENERGY-INTELLIGENCE, the (OR EIN SOF-אור אין סוף)the Endless Light. It is trough this energy that our soul comes from. We are all able to reach for It if we really want to and we follow what I mentioned in the chapter of the Endless-Light.

One thing is shure is that the world we perceive is an illusion.in order to perform what we call miracle,like healing people,make things happen like suddenly bring an elephant from nowhere,or more so rise a dead person.

During all our rich history,many are the prophets and sages that did just that. This great people had the know how,the mind power and could directly draw CHOCHMA ILAAH- חכמה עלאה (highest intelligence) from the unknowable and most daunted of all the Mighty CREATOR.

This of course is the KABBALAH. Where all secrets of creation is revealed. Those are the jewels the ALL MIGHTY gave us,to draw from the highest of powers of the Universe.

But again this powers were given to peple to use very wisely,I say wisely because every Miracle that this person will perform will have never,I say never to take credit for it,he has always to recognise the Creator as being the source and them being the vessel for the person needing the miracle.

If it was not so,the person performing the miracle will draw from his account,meaning from his health if it was the case,or years of his life and so on. There is another way of doing miracles,trough parchemin writings or in semi-precious or precious metals.

The writings or engravings have to be done after fasting,cleansing of the mind,away from lust or gain of notority. Then you have to know the days of the month and the hours of the day that you should do the above mentioned items.

A third way of making wonders is hypnosis,like I said in previous chapters,since faith does miracles it is because the mind is the ultimate tool to heal the person by guiding him/her to reverse an illness probably

Most of them,including Cancer. Medicine or science will not agree with me but they can't explain a miracle. For your knowledge Medicine and science in general are not a perfect skill but only through observation and suppositions,they are true for today but archaique for tommorow.

Television would have been science-fiction a century ago.but today it is reality.the same for a million things. We have entered the Messianic era,the Aquarius era. We entered the era that the world will discover the CREATOR'S greatness. All this knowledge is drawn from HIM.

If you want to enter the the world of the unknown,it is time to begin.

My book saphire & Diamond is very deep in this subject,you will learn the language of angels,of the letters,nubers,planets etc....

Below are a sample of some pages of my book.Hebrew is a must. To be a Kabbalist you have to start now the time is short and the task is enormous and hard.

האל״ף בי״ת וסמני הרזים

אלף ב״ת עברי		סמני צבא מרום		מלאכים	
א	ס				
ב	ע				
ג	פ				
ד	צ				
ה	ק				
ו	ר				
ז	ש				
ח	ת				
ט	סופיים				
י	ך				
כ	ם				
ל	ן				
מ	ף				
נ	ץ				

THESE ARE THE ALPHABETS FOR THE LANGUAGE USED BY THE ANGELS AND THE ARMIES OF HEAVEN,THE PLANETS.

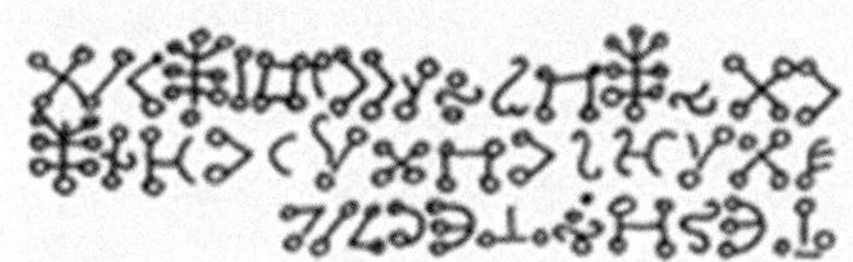

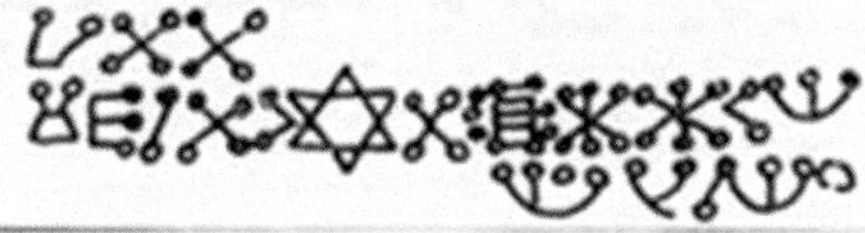

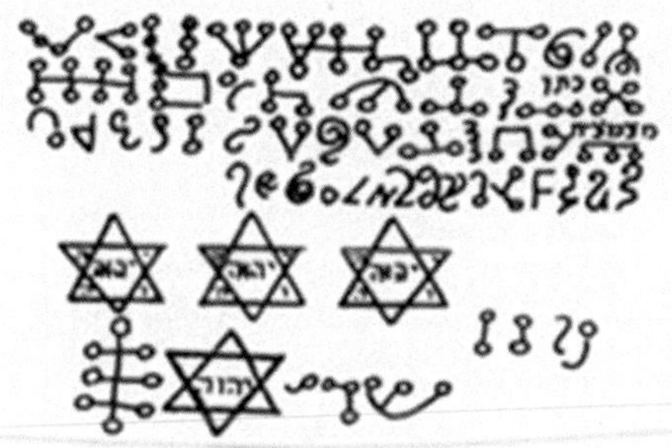

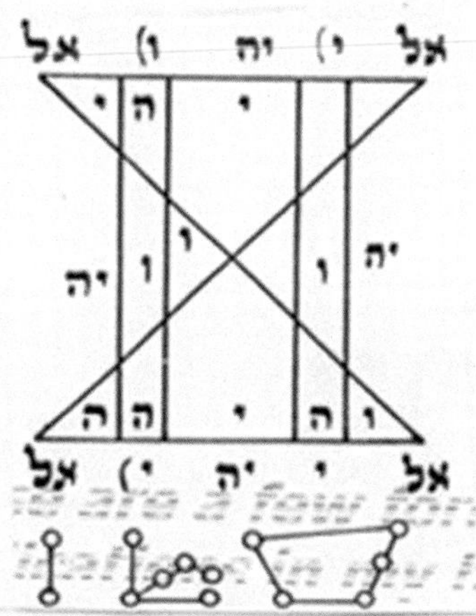

These are few of the formulas that have different applications from my book "SAPHIRE& DIAMOND". Each one is a recipe for a miracle as we call it,of course needing to know all the surrounding requirements.

GENESIS
In the
BEGINNING

בראשית

Chapter VII

Genesis- בראשית

"In the beginning G-d created heaven and Earth. The Earth was empty and without form, there was darkness on the face of the depths, but G-d's spirit moved on the water's surface." (Genesis).

In Kabbalistic terminology, THE BODY is called Earth, given the fact that we are made from dust. It is a mass formed from Men's semen penetrating the Woman's womb. The moment it encounters the ovum it creates a living "Mass without form.

On the other end of the spectrum, "Darkness on the face of the depths". In Hebrew, "TEHOM" תהום'means Depth and inverting the hebrew letters, would be "HAMAVET המות(means Death), in the book of Genesis. The aura does not abandon the BODY, but follows wherever the BODY, is taken. Once in the tomb, it finds itself alone in darkness, "in the face of the depths", because he thinks that is what's left (from who he was), an inert Body, decaying and all his life passes before him, and he thinks that it is the end,TOTAL DARKNESS.

"But G-d's spirit moved on the waters surface" (Genesis). The bone of resurrection, called in Hebrew "LUZ זלו (Backbone), that does not decay at any time. It is called in Kabbalistic terminology, water, and the spirit of G-d is our own energy-intelligence due to the fact that is coming from the Creator.

G-d said, "There shall be light, and light came into existence." G-d saw that the light was good. (Genesis).

G-d said, "There shall be light, then the light or energy intelligence comes out from the tomb and sees the light, the light of the Creator (and the light came into existence) because in Hebrew a cemetery is called "Beit CHaim'בית חיים (the house of the living). Life starts at death, and not at birth, contrary to our understanding.Those who are born, they are destined to die, those who die are destined to be reborn(ethics of the fathers).

When we are in a spiritual form, or energy-intelligence form, there is no more sorrow, pain, or worries (if at death he was a good person; see chapter on reincarnation). "G-d saw that the light was good", meaning that this soul or energy is part of G-d. Once liberated from the shell that is the Body (Earth) it goes through a process of acceptance or rejection by the Creator, that will be explained in more details in subsequent chapters, "G-d saw that it was good."

G-d said, "Let us make a man with our image and likeness," (Genesis). Signifying, the Creator with the departed soul or energy-intelligence of the deceased takes the Mass, being formed in the womb of the third party, "the Woman".

The energy intelligence, or soul is attracted toward the new life, a Mass without form (surrounding light "OR MAKIF") אור מקיף surrounds it, and gives instructions according to the soul's energy- intelligence capacity, or parity,what ever is left after departing from a prior life, to take a human form in the womb. An embryo is being formed to a human likeness of G-d, of energy-intelligenceor our soul.

Here. again when G-D says let-us do. He also refers to the four components AIR,FIRE,EARTH AND WATER.אויר אש מים ועפר All living organism in our world is formed from these components,.So G-D orders the latters to form by HIS request. The last and most important component is G-D's energy intelligence or, our soul

"Birds of the sky, the livestock animals and all the Earth" (Genesis); it means that, depending on the purity of the soul, if it was beyond a certain purity of positive energy-intelligence, it will reincarnate as a human, if below average, it will come back as a lower form of intelligence, like birds, animals or Earth, etc. The highest form of intelligence being human, then animals or birds, plants and finally Earth or minerals.

G-d created man with His image. In the image of G-d, he created him." (Genesis). It is repeated twice, again to the hidden meaning that one is in physical form, and the other being the energy-intelligence, forming the living mass in the womb, as a human. "Male and female He created them," the Soul's energies are divided as follows: Masculine, feminine, positive, and negative. On the physical plain, the soul or energy intelligence will take, the form of man or woman depending on the Tikun תיקון(correction) Karma, as it is called by other cultures..

G-d said "behold I have given you every seed-bearing plant on the face of the earth."(Genisis).He mentions another form of intelligence-energy,thet would be additive to our Energy that maintains the Body and Soul together, like water to a plant. The body will die without this type of energy additive that maintains the Energy-Intelligence together with the Body, that is called in kabbalistic terminology, "the Shell."

G-d blessed the Seventh Day and He declared it to be Holy," (Genesis). The Bible mentions that G-d created the world in six days and rested on the 7th." (Genesis). ' G.d's day in esoteric explanation, is that every day is one thousand years of ours. Within the Hebrew calendar we count from Adam's birth. We are in the year 5752, since we are counting from a lunar calendar, the next day starts at sundown. All that this means is that we are already in the sixth day, due to the fact that 5750 is 5 and ¾. Since the next day starts at sundown, or at 3/4 of the day. So 5751 will be the beginning of the next day or Messianic era,the FRIDAY of creation an d so on.

The seventh day (G-d ceased from all the work that he had been doing). Denoting the end of the cycle of current life, of a bodily form. It is understood from a quote by Kohelet קהלת ("King Solomon") that states, "There is nothing new under the Sun", meaning that there was life before the beginning of Adam. Life endures 6000 years per cycle and one thousand years of desolation of bodily form, but because we humans are our own destruction, and our own worst enemy.

Today's situation will explain this statement, (Ozone layer depletion, contamination, over population, wars, especially nuclear threats). "And G-d rested on the seventh day" (Genesis). This rule goes on for all creation as the human body needs a Sabbath day שבת to rest, like a machine that works non-stop will fall apart.

A Sabbatical year for the Earth, is every seventh year, we have to stop planting and working the earth, and let it rest. and if we don't; the fruits and vegetables will not be as good as it use to be, and the land will decay until it becomes a desert. We do have proof from history of a lot of land, that has become desert with no logical or scientific explanation.

Every creation has to have a rest, and the number 7 comes back and forth in the Bible, and we will elaborate in another chapter.

All earth is filled with G-D's essence,nothing can exist without His essence. We will have a longer explanaition further in the chapter "Endless Light"אור אין סוף

REINCARNATION
LIFE AFTER LIFE
ENDLESS LIGHT

גילגולים חיים אחרי החיים
אור אין סוף

CHAPTER VIII

Reincarnation, Life after Life, "Endless Light" גילגולים- חיים אחרי החיים אור אין סוף -

The Endless Light is credited to the Creator's resplendence, not light as we conceive it, but energy-intelligence in its pure form. The Creator is all pure and from His essence all galaxies, planets, stars, and all life form as we conceive it are created.

Nothing comes to being without His essence, (The Universe is filled with His Glory). We are created in His image, which is the meaning of his Energy-Intelligence or essence. Our soul is only a spark emanating from the Creator. We live in a four (4) dimensional world, that is called in kabbalistic terminology with the Hebrew abbreviation of:" ABYA"אביע

A –א is for Atsilut אצילות(Emanation)in Hebrew the letter E or A are written with hAleph

B-ב is for Briah בריאה(Creation)

Y-י is for Yetzirahיצירה (formation)

A-ע is for Assiah עשיה (Performance or deed)

See Figure One and three

FIGURE No.1: The 4 Parralel Worlds or the 4 Dimensions.Each dimension is divided into 10 levels of Awerness and then each level is again subdivided to 10 m0re and again to infinity.

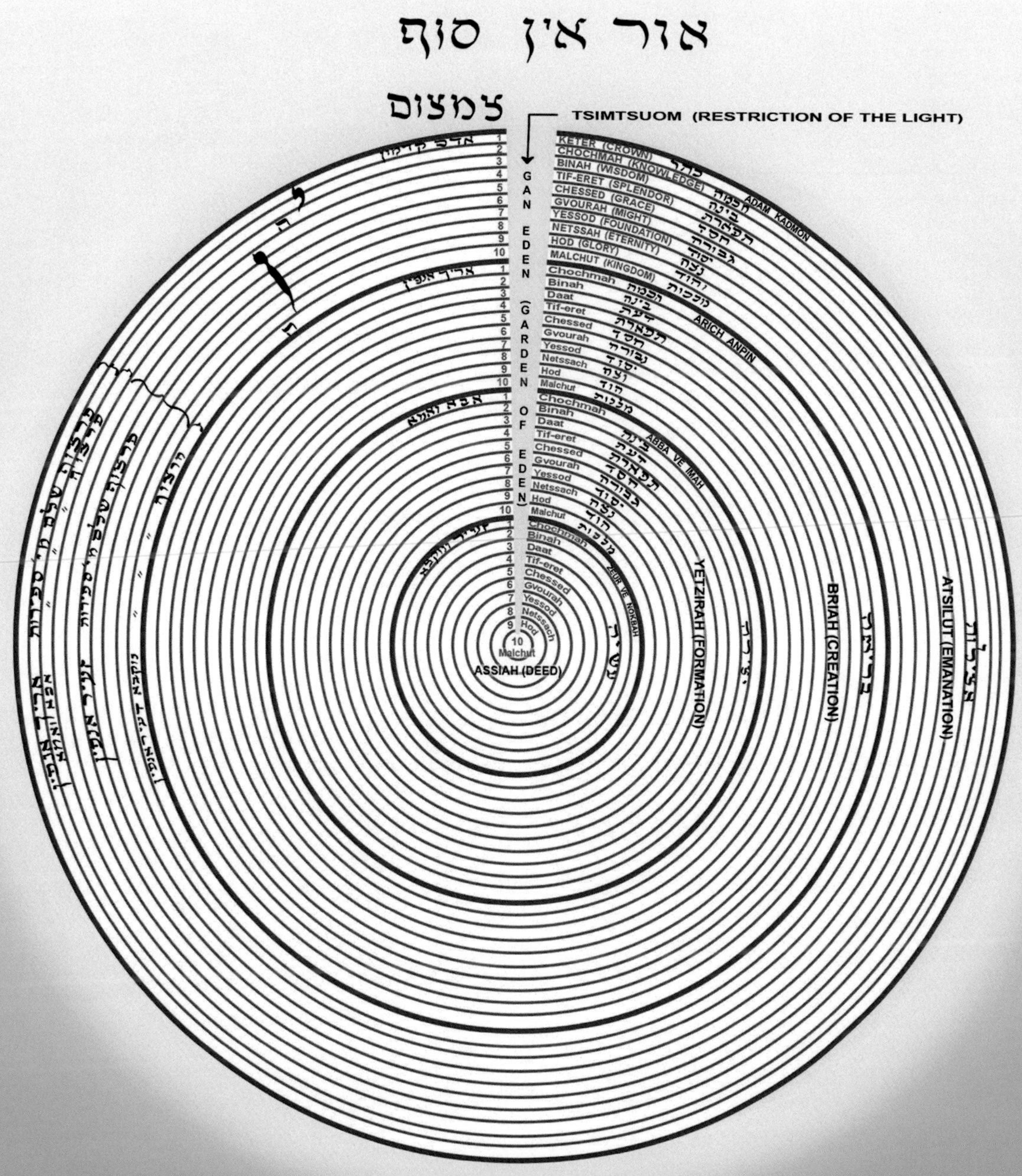

FIGURE No.2: The Four Dimensions of the Earth. We can appreciate a section of the Earth. It shows the four sections that equals the four dimensions, the core being the PURGATORY.

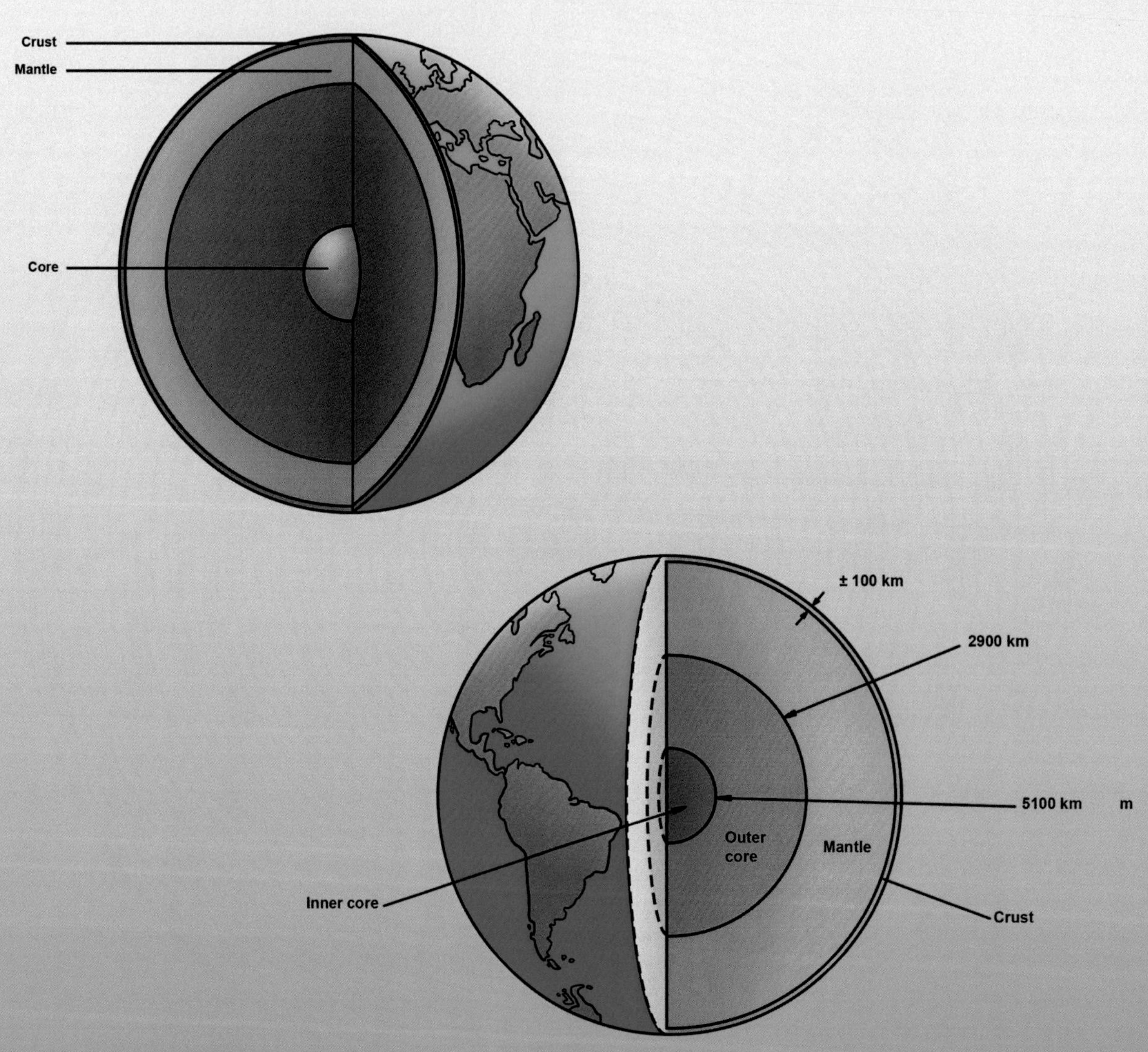

FIGURE No.3: Our Physical & Spiritual World.

The 10 levels of awareness of our world and the 10 inner levels of negativity awareness in the UNDERWORLD toward the PURGATORY,being the 4 sections of the EARTH.

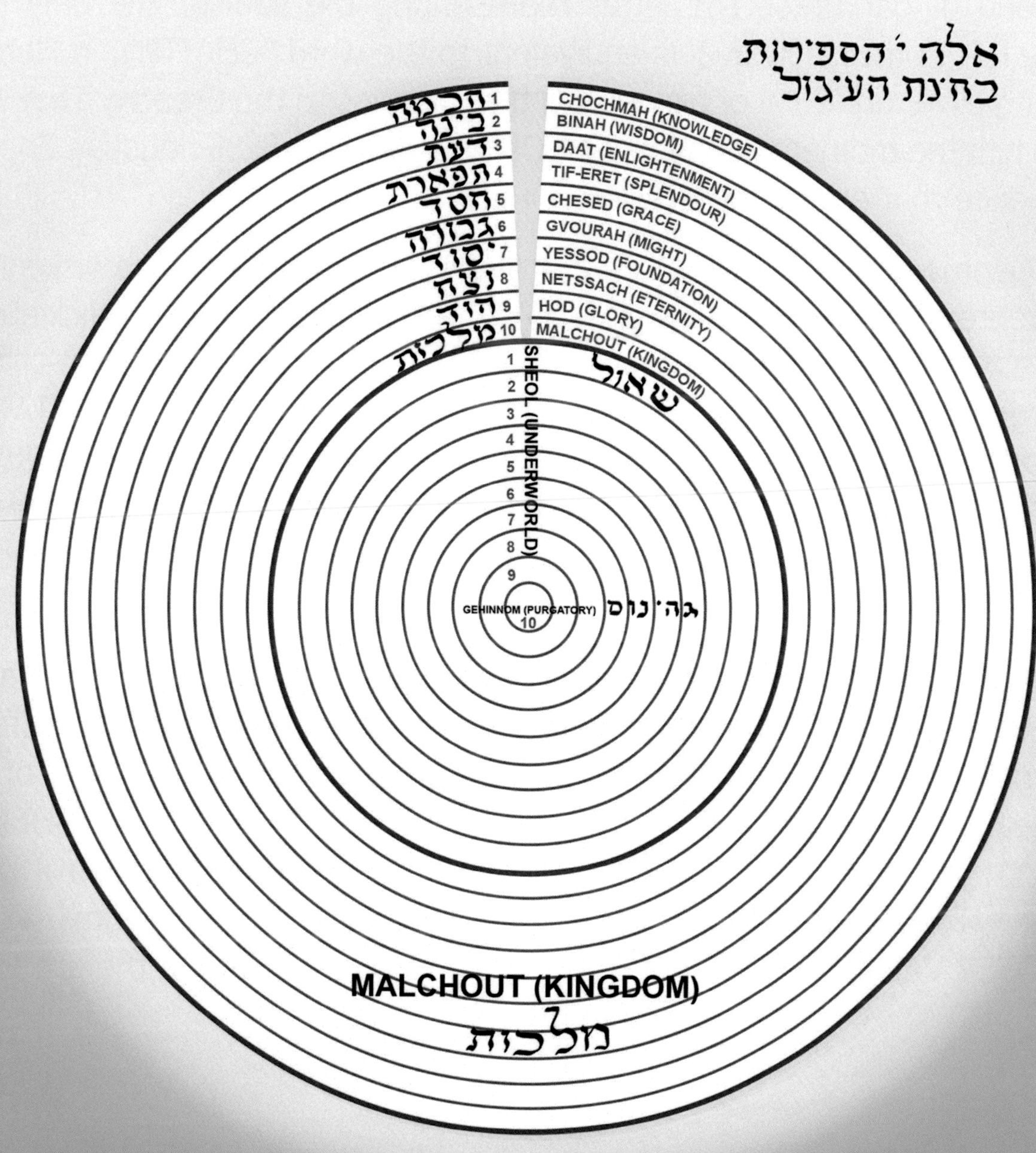

Also in the underworld, the World of darkness, there are 4 levels of awareness. See figure 2, and it's called SHEOL שאל. A). SHIN(SH) ש, it is the first letter SHADAI-שדי that means water, meaning that humans entering the world of darkness, due to impurity, no saint or pure entity energy enters the world of darkness. The letter representing the soul or the only pure energy left to that individual is equivalent to the word "CHAI" חי, in Hebrew meaning alive, or " 18% of his energy, that belongs to the Creator. That 18% leaves him the moment he enters SHEOL שאל, making 82% negative Energy-intelligence that enters the UNDERWORLD.

2) (Aleph) א or the "E" of SHEOL שאל = AVEN אבן in Hebrew means wickedness as he enters the ABUL-אבול (gateway) of the kingdom of evil. Wickedness Afela-אפלה (darkness). The one he chose, the kingdom for Elilim-אללים, (false gods). He becomes ABOUTZ -אבוץ(galvanized) EM KOL CHATAT-אם כל חטת(the root of all evil), AMOSH-אמוש, (DARK) he becomes ENERGIAT ACHPAR-אנרגית אבפר (fossil energy) oil in other words. AVICK -אביק(powder/dust) there he arrives at the EZOR SHIPUT –אזור שיפוט(area of jurisdiction of the VAV.-ו

3) (0) or Vav in hebrew. =. He enters the level VULKANI-וולקני (volcanic). VAY MIBETA –ווי מבתה(in Aramaic) Woe from within. VAZATYA-ו זיתה (fools) for arriving at this level due to their foolish behavior in the light of the line,meaning a life span of a human it becomes VUSAT ALYEDE SHEDIM-ווסת על ידי שדים (governed by demons) or negative pushing him toward VERIDOUT hadelek-ורידות הדלק, (viscosity of the oil petroleum). VITER ELCHOSER VADAOUT וויתר אל חוסר וודאות– to uncertainty in the attraction

toward his. final station, depending on his level impurity). He could be stored as oil or be attracted further down toward LAMED) L-ל

4)(LAMED)-ל or L= world of LAVA, GEHINOM-גהינום (PURGATORY), there he is UKHAL-עוקל (consumed).[To every creation there is Action in the four above mentioned dimensions;(see figure No. 1). In every dimension there are Ten different levels of awareness, called Sefirot-ספירות.(See figure No. 3). Every Sefira-ספירה (sphere) is again sub-divided to ten more with the same sub-definition to infinity. The pipe or stream of light (awareness or intelligence) that flows from the Endless Light, that comes to illuminate the entire universe. only a small amount can a person handle. Like a person who is in a dark place for a long time,then comes out and tries to look at the sun. It would be like trying to feed 10,000 watts to 100 watts bulb, the bulbs will explode.

Similarly we have a possibility to perceive only a very limited amount of light of the Creator. This is the fact that we are limited by the curtain of darkness or negativity, called the restriction of the light.צמצום אור האין סוף

Explaining more in detail, our brain is limited to approximately a 5 to 10% level of understanding. We are limited due to materialistic programs that we feed our brain with Consumed in the fire of Purgatory, center core of the earth or Gehinom-גהינום (the material purgatory). The last letter of Gehinom is M-מ for MAGMA.

There are two names SHEOL-שאל or GEHINOM-גהינום and they are both referenced in the four different lower levels of negative awareness. Each level is then subdivided unto ten levels of awareness and again to the infinite.

Greediness, pride", anger and sadness. All this impedes us to expand and reach a higher level of awareness and intelligence. Likened to a computer if you put a program for accounting you will have only accounting that can come to the screen and if,let us say you put another program for games, only games will appear.

Also our brain is similar to a computer, we only think about money, more and more, to have a beautiful house, but when you have it, you get tired of it and then you want a better one, to endless ambitions. So there is no time for anything spiritual, or to reach higher levels of understanding, because to do so, you have to alleviate the memory bank, delete a lot of data such as: anger, vengeance, pride, sadness, etc. and insert the correct program for developing the other 90% of unused intelligence.

This program is called the Kabbalah. To cite an example, a scientist is portrayed with long uncombed hair, badly dressed not worried about his appearance, sometimes he forgets to eat and he does not care what people think of him. Sometimes he looks very distant, that gives you an idea that his mind is not occupied with materialistic thoughts.

He knows how to concentrate in the direction of research. He seems to be in a trance.

Now it is not necessary to eliminateeverything of wordly matters, but to do the necessary minimum, because we live in a society and we cannot just ignore that fact. Let's just for argument's sake discuss food. We eat to live, and not live to eat, as many people do, if we only understood the principle of life (we will expand further in the appropriate chapter). In the physical world, when we are in our mother's womb, we are connected through the umbilical cord, also in the metaphysical or spiritual world. We are connected through this tube that comes through the four different dimensions. From the Endless Light of the Creator,(see fig.3) He feeds us like the umbilical cord. Without it, a baby will be born dead.

"G-d took the man and placed him in the Garden of Eden to work it and water it. G- d gave Adam-אדם one Commandment, saying you may definitely eat from every tree of the Garden, but from the tree of knowledge of good and evil; do not eat, for on the day you eat from it, you will die." (Genesis). In this case the tree of knowledge is the tree of the Four Branches (A.B.Y.A.)-אביע or to lower oneself to a lower level of awareness, to the level of ASSIAH-עשיה, (lower of the worlds). He will surely die (because anything material has a beginning and an end) likened to the life of the line that has a beginning and an end. The serpent was the most cunning of all the wild beast that G-d made. (Genesis) The serpent is portrayed as evil or negative influence that one carries within, or the "S" for Serpent or "S" for Satan -שטןthe devil within us.

"The woman saw that the fruit of the tree was good to eat and desirable to the eyes and that the tree was attractive," (Genesis). As a means to gain intelligence, we have sometimes bad ideas coming to us disguised as good ones. Our other side (SITRA ACHRA)-סיתרא אחרא, or our beast side, tries to convince us that it is the right thing to do, even though it is wrong. The other side is strong and tempting, especially in the world we live in, like a cigarette, we know that it is bad, and it causes cancer, you pay for it with your own money that in turn destroys you. In other words, the Serpent.

And she took some of its fruit and ate it, she also gave some to her husband and he ate it" (Genesis). denoting that if we do bad and to feel more sure that we did good we convince someone others, in other words to share a hidden guilt.

"The eyes of both of them were opened and they realized that they were naked." (Genesis). They both were in a lower level of consciousness, due to committing a sin that brought them to a lower dimension, where they found themselves naked (of body).Only in a soul's form, in a physical world,the world of Assiah."G-d made skin garments for Adam and his wife, Eve. He clothed them." (Genesis). The skin garments were their new bodies to their soul or energy-intelligence. Up to now, I have only showed one of the seventy meanings hidden in this chapter of Genesis, so imagine how many more secrets are in the Bible to be discovered, especially, trying to understanding our physical body and its hidden relation to our invisible soul.

In this present world, Adam was the first man, and all humanity descends from him. The latter being a spark of the Creator and from that spark, it was subdivided into two sparks; one male and one female. And the man knew his wife Eve" (Genesis). Adam comes from the Hebrew word Adama- אדמהEarth. Hava -חוה(Eve), because she was the mother of all humans. In the Bible it mentions "knew", in this case, it meant the act of procreating.

"They had Cain and Abel and two daughters that were both wives for Cain and Abel and so on. So we all descended from Adam and he in turn from the Creator.

Also ADAM-—אדם means's men in general.So we all together form that man or ADAM,and EVE. WE have them also divided into five Billion people all of us having a masculin and a feminine side, thus, the two possibilities are :

A)-if we can achieve all the above mentioned separation still being in a bodyform,to reincorporate that man or woman i.e.ADAM and EVE,would mean that all masculin should separate to form ADAM, as well the feminine to form EVE.Then negative to form the Serpent and all positive would join G-D Himself as He is all pure, we then will be able to return to Eden. Thus the two possibilities are:

1)- if we can achieve all the above mentioned separation still being in a body form, the ISAIAH'S profecie " And it shall come to pass,that before they call,I will answer;and while they are tet speaking,I will hear. The wolf and the lamb shall feed together,and the lion shall eat straw like the ox:and dust shall be the serpent's food. ie. The MESSIAH's arrival . Destruction of all flesh and material form.Then evry energy will join its own kind. As I wrote in the beginning of this chapter, life starts at the death of the individual.

Let us review in more details the life form process: When a person dies, his Energy- Intelligence remains with the body for a certain period of time depending on his degree of purity or negativity. If the person was very pious and pure, especially if he studied Kabbalah and Torah-תורה in his life term on Earth, he will not be attached to his physical body. Earth is the Antechamber to Paradise or Purgatory (both will be explained in detail). So, if a person studies Kabbalah during his life term, he has the knowledge to shorten his attachment to his decaying body, and knows the process of reincarnation.

Let's assume that his Energy-Intelligence is for example 2000 or 90% pure, (those numbers are figurative only), his soul or Energy-Intelligence reaches higher levels of awareness to the 2nd or 3rd dimension. Like fire, it reaches toward the sky, but; dust (ashes) falls back to Earth. To actually reincarnate, depending on our Earthly accomplishments.

The purpose of reincarnation is to better and purify the energy intelligence of the soul when this 90% pure reaches certain heights, it encounters a purer form of energy. When trying to incorporate and unify with "The Endless Light of the"Creator", it is trying, for example, to connect a 100 volt cable with 1 0,000 volts, a spark will occur rejecting contact.

The soul or Energy Intelligence will circle around its level of awareness untill it is attracted to a new life form, and the same is for the lower percentage forms of Energy-Intelligence. Reincarnation starts as follows: A Man and a Woman are doing the act of procreating. Depending on the quality of that moment, will be the purity of the soul that is going to be attracted. Like turning a radio on, depending on the channel that is going to be tuned in, will be the channel that you are going to hear.

The same thing applies with humans, the moment of the act is very important to have; 1) pure thoughts, the way love is made at that moment, in

Hebrew man and wife are called ISH- איש and ISHA אישה written with Hebrew ISH-איש(man)ALEPH-YUD-SHIN and ISHA-אישה(woman) with Aleph-Shin-Hey. The man has Yud –י and the woman a Hay-ה, those two letters forming Y-H –יה or the name of G-D . Taking those two off, when making love, is taking Godliness out of the love between the man and the woman, leaving only ESH-אש(fire). That is what will be left of their love, fire to be consumed without essence for their love to survive, or the Y-H יה G-d's essence.

2) The best day for procreation is Saturday (for reasons I will explain further) 3) It is very important that the woman should be clean at least for seven days, following menstruation. It is very dangerous for the new born, (see figure 4).

FIGURE No.4: A Human Embryo,a Brain and an Urchin Egg Surrounded by Sperm.

A human Embryo in his 4 dimensional worlds. The 4 different parts of the Brain. The sperm trying to penetrate the egg searching for life.

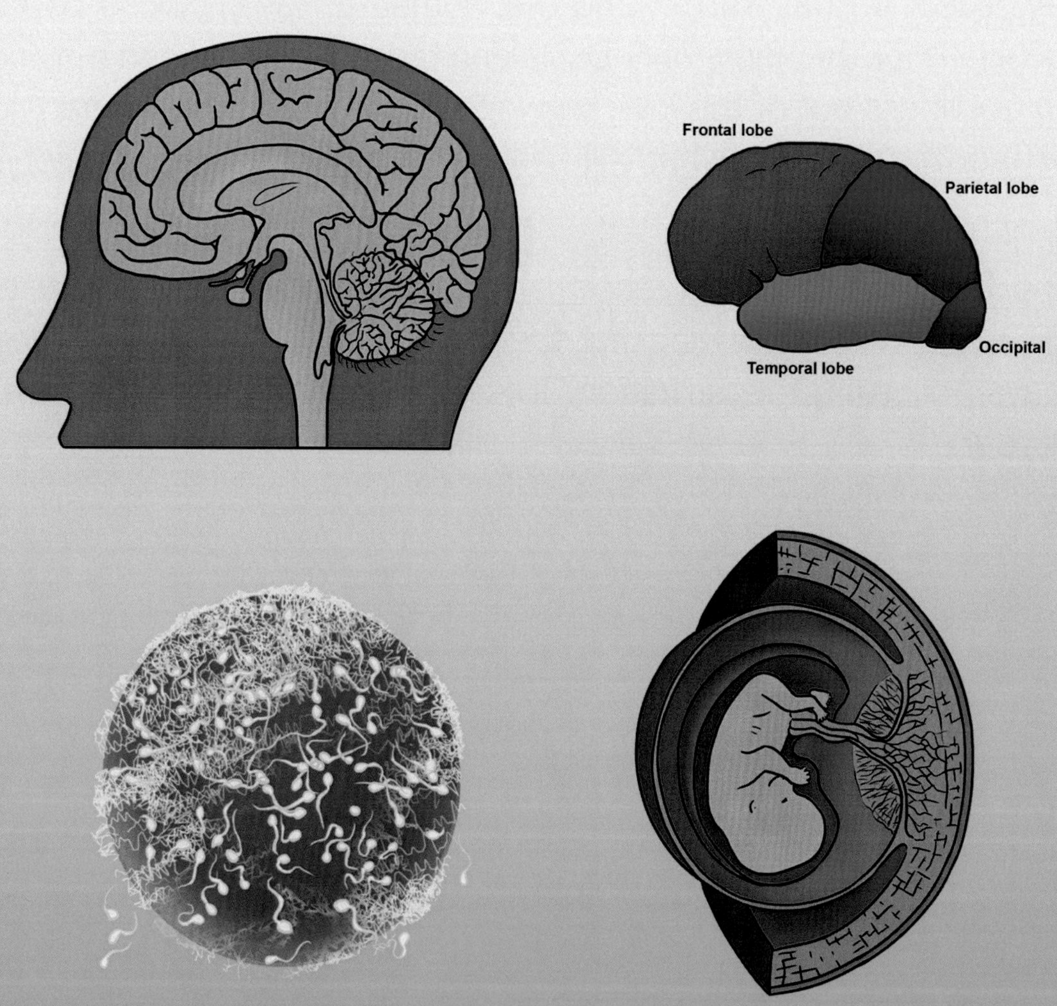

The woman after the menstruation period becomes fertile, and we know that a woman during this period goes to an unclean seven days, and could transfer to the child a perturbed energy that could cause the embryo to be born with a Down syndrome or a deformed child, as children that have progeria, that age prematurely. It is caused by the lack of coordination between the surrounding energy-intelligence (OR MAKIF-אור מקיף)and the inner light energy (OR PNIMI-אור פנימי) "the mass without form'so one of the functions is being accelerated, "the Biological clock".in order to change the course of events Hypnosis could be a possibility to try and reverse the process.

Under hypnosis,you bring the child back to the conception stage and give to the child the actual instructions of normal growth, how to develop this connection directly to the subconscious part of the brain, plus other treatments after the hypnosis. The process will be reversed and stopped up to that point, with G-d's grace. Through hypnosis,it is possible to project a cure for a multitude of illnesses. It's also possible using the same vehicle of hypnosis, to effect a cure for different types of cancers. This can be done in a variety of scientifically accepted techniques.

Photo representations or video, can on a specific progress of a particular illness or cancer, be projected and quite accurately prognoses -Furthermore, these methods can enlighten the part of the body affected. The photo representations, or video illustrates the multiplied nature of the bacteria as well as how it proliferates in its ravages against the relatively unprotected host. The patient so afflicted must see these pictures or video at least in twelve separate occasions. On the following day he is expected to return in a state of fasting.

The hypnotist and surgeon then proceed to lead the patient in a sonombulistic trance to his visions of lesions, as he had watched so many times. They also must make the patient realize fully, that what he has seen is the condition as it exists in his body. They further explain the intricacies of the area.

The following stage is to have the patient become cognizant of the fact that he, the patient, is the sole driver of the antibody action phenomenon. making him feel that he is a super human and reminding him what he saw on video and that he can now control the red or white cells and destroy the cancer or other alien infections or parasites and control the levels of the red and white cells in case of Leukemia

He is further taught how to surround, attack and finally eliminate both the cancer cells and any accompanying infections. Because one's mind can wondrously affect bodily functions it can, in this case, perform what no other technique can. In this manner, every form of illness or turpitude can be either corrected or completely cured. For this reason, we are apt to often conclude that faith can perform wonders bordering on the miracles. Because of our energy-intelligence, the body so stimulated by mental means that can eradicate and fully eliminate any and all conditions harmful to good health.

The key to the scenario is the know how; the ability to control our minds over bodily functions and specific conditions. The babies life is formed in the woman's womb, the woman gives an energy of, let's assume, 750 and the man of 1 000. So it will be a mass requiring a reincarnation of 1750 positive divided by 2 = 875 average purity of the soul or Energy-Intelligence, so those higher Energy-Intelligence are really very difficult to be attracted back to this life.

The lower ones have more possibility to be recalled, because not every day is there a genius or saint born. In addition, to our sorrow, Bad and Evil are overwhelming. Coming back to our process, that 1750/2 = 875 average purity of energy intelligence has the knowledge to incarnate a human form. But if the baby was conceived in a period that the woman had perturbed energy because of post menstruation, or drugs, or rage, etc. the energy of the ovum will be perturbed and will not form properly, due to the fact, that the energy Intelligence or soul will send the information, but because the perturbed energy will receive wrong signals, the receiver being defective and from that fact it could form a mongoloid, or a baby with one arm or two legs joined together or two babies sharing a common head (G-d forbid).

Now in the event that another person died and was a bad person, he could be reincarnated not in a human form, but in other forms of intelligence.

The 2nd Energy- Intelligence level, after human, is an animal intelligence form, that is a lower intelligence than human, but still can move, eat, and is influenced by the sun, rain, wind, etc...

Then comes plants in general, that do not move by themselves and do not make decisions, but are influenced by the sun wind and water.

The fourth form of intelligence, is mineral, which is not influenced by sun and water, etc. It is known that all of the above are composed of atoms, and every atom is composed of protons, neutrons, and electrons. Like billions of solar systems from every entity (see figure 8).

Everything has a level of awareness, and a level of Energy-Intelligence. If one lowers oneself during his lifetime to lower than 5 1 % positive, there is no return, because after 50% positive turns to 50% negative and the more negative one is, the lower the Energy-Intelligence.

The reason the Israelites were brought immediately out of Egypt, is because they were in Egypt under slavery, and a great majority of them adopted Egyptian customs, like idolatry and their level of purity of energy-intelligence were at 51% and one more degree of negativity would have made it impossible for Moses to save them from the world of SHEOL. 51 % Positive and one more degree of negativity would have been the last, with no possible return, because even having a high degree of purity, it is hard to deny temptations, with more negative than positive it's impossible. But don't misunderstand me, to be lower than 51%, one must be very bad, not just denying the Creator, but to destroy G-d's creation as killing or destroying for fun and feeling good about it... that's bad!

Now when on the two opposites, we have the Garden of Eden, or the Purgatory, (see fig. #1). Let us expand on this subject. The spiritual Garden of Eden is in the Endless Light. Thus forming unity with the Creator.

I personally had N.D.E.(near death experience) in an accident that happened in Eilat,Israel. I was I 9 years old, and was in the army in Israel. I was with some friends during a few days vacation from the army. We had decided to travel to Eilat, a city, south of Israel, in the sea of Reeds (the Red Sea).Further south there is an island bordering the Sinai-סיני called EE-HAALMOGUIM –אי-

האלמוגים or the Algae Island. At first we started playing in the water, male and female soldiers.

After a while, I started swimming toward the isle located about a mile from shore, where we were. The current was very strong. At my age then, I was not concerned with the possible dangers. I thought nothing could happen to me! When I was a few hundred yards of the island, I felt a pinch in my calf. Due to the pain, I made a sudden move and cramped my leg.

The pain was unbearable. I couldn't swim any longer. The current was very strong. I was tiring fast from trying to keep my head above the water, but the big waves made it impossible. I couldn't control the intake of water, and started drowning. I was sinking fast and could barely see light above me anymore. I was panicking until my strength abandoned me.

I was holding my throat, suffocating until I became Suddenly, in a state of awareness. I felt good and light. The water was there, but I didn't feel its wetness. I was very calm. I looked under me and saw my inanimate body, holding its throat but I didn't feel strange, nor cared for my sinking body. below me.

It was then that I saw a bright light. I felt pulled up to the surface, toward that light. I then was out of the water and kept going up in the sky. The strange thing was, that I felt myself as part of the wind, as if I were transparent. I experienced all sorts of visions, but without feeling any emotions.

All my life went by me, as if I was watching a movie, my life in a re-run. All my actions were in revue. I felt a certain presence with me, but I did not see anyone. The next vision was my family, parents and brothers dressed in black mourning, following my casket, covered also in a black cloth with a star of David in its center, layered in gold thread. I did not feel any emotions whatsoever. Getting a closer look at my parents faces, tortured by the pain of my loss.

At that same moment, I felt a pinch in my heart, suddenly I felt pulled down toward the sea. I then submerged, approaching at high speed my inanimate body that was in the depth of the sea. I was now in my body, fighting the storming waters, swimming upward with a sudden found strength. I could barely see light above me but it grew rapidly.

This brief moment seemed like an eternity, until I broke through the surface, gasping for breath, coughing and vomiting. Not far from me was one of my friends who saw me and grabbed me, until the others came and got me out. Jugging by the distance they were at least half an hour from me when my trouble started.

Yes, my dear readers, l did die. The only feeling that made me return to this life was the suffering of my parents. I did not care for my body. I return because I was still attached to Earthly emotions (my parents).

From this, you can understand 'AKEDAT YISTCHAK'-עקדת יצחק the sacrifice of Isaac-יצחק by Abraham-אברהם. The latter was in a higher plane of awareness. In that level he did not feel as we do in this material world. He followed G- D's instructions without any Earthly feelings, as we are bound by them, as I was. Abraham was confined by the limits of his body, which I was not.

This showed Abraham's greatness I will explain in the following chapters how can this happen. I knew from that moment on, I was then 19 years old, that I wanted to learn what's beyond comprehension. Of course I do not know everything, but today I am 39 years old and I have learned a lot, and there is still a lot to learn. We die learning.

Coming back to our subject, Paradise is within higher levels of consciousness, the higher you go, the higher peace you find, Paradise is unifying with the Creator's oneness. Now, the other end of the spectrum is Purgatory. How does it work? When you descend to a lower lever of negativity, you are turned into one of the three (3) lower levels of intelligence form, but you are still in a level of awareness, understanding your existence.

If you are in a real low level of negativity. for example: 70% of you becomes fuel. and as we know fuel comes from deposits of fossils, human or animal and since the negativity attaches to material and not spiritual, it remains with the body, then it is converted to fuel, and still feels earthly emotions, depending on its level of awareness. It is closer to the earth's shell the more he understands his position toward life in human form, and the more he intensifies within the center of the depths of the Earth, he starts to be more aware of purgatory. "The Lava" fire that burns the souls of evil, as we understand it, that's purgatory.

We paint it abstractly as a demon with a tail, horns, etc. So we have a chance in our life term to decide to act positively or negatively, we are the third column that draws from positive or negative, we draw to our decision from one or the other, as it is mentioned in the Bible. (Exodus 14:39).

"G-d went before them by day with a pillar of clouds to guide them along the way, by night it appeared as a pillar of fire providing them with light. The pillar of clouds by day and the pillar of fire by night and never left the people. Here, we see the 3 column system, the negative (clouds), the positive (fire), and the Israelites as the 3rd column. As the 3rd column we can choose day or night, clouds or fire, darkness (night) or light (day), evil or good. In the back, are the Egytians (Evil), to the front, the way to the Holy land (Good). Being the middle, we are in a position to draw and choose from the two, and have a deeper understanding.

In the same way the atom is composed of protons (fire), neutrons (clouds), and electrons (Israelites). There are other levels of awareness to understand this message. A column of fire lead the Israelites, they were in the middle, and behind the column of smoke, and that forms the 3 column system that the Kabbalah goes by.

We, in this life time can upgrade or degrade our Energy-Intelligence, as a Parable states. There was once a king who had three sons. This king decreed that anyone in his kingdom who wanted to gain a position in the hierarchy had to gain it in deeds and not by buying it with money. So when his sons were 18, 19, and 20 respectively, the king said to them, "I decreed in my kingdom and this also applies to all of you. For there is a ship leaving for an island, it will drop you there. In six months, it will pick you up. Use this time to the best of your abilities to gain for yourself a position in my kingdom.

"So it was, they were left on this island. The first day, they saw 3 gates to 3 gardens, one with all the food in the world, the other one with all the wealth in the world, and the third one with all the beauty of the world. There was a sign on every gate than when you leave you cannot take anything with you; you come empty, you leave empty.

So the eldest decided for the gate with all the food, he entered and immediately began eating day and night, and fell asleep afterwards. The second son went for the wealth. The moment he entered he found diamonds, he started picking them up and filled his pockets, then he took his shirt off and made it like a bag and filled it with all kinds of precious stones and then the pants followed and pulled everything behind him .

The youngest son decided for the beauty. He entered and started walking and he heard a beautiful song of the birds, he tried to listen and understand what they were singing and then he saw a beautiful glow in the river. He tried to follow and discovered where this beautiful light originated, so he kept following it Weeks passed, the oldest brother became sick of eating, fell ill and died. The second was naked and freezing, because all his clothes were filled of precious stones. When the time came to leave, he came to the gate, the guards confiscated everything he managed to gather and beat him out of the garden showing him the sign that reminded him that he could not take anything with him the moment he left.

The youngest son learned many things such as beautiful bird songs, and many more lessons of the beauty of creation. He was happy. He met his brother naked and sick, they boarded the ship back to their father's kingdom and once they arrived, all the citizens were there to meet them. When they came down everyone came to greet the youngest one. The other no one paid attention to, due to his nakedness when he tried to reach the palace, he was beaten by the populace, even when he mentioned that he was the king's son, everyone laughed at him.

When the youngest one met his father, he told him how much knowledge he had acquired and that the guards could not take that away from him, explaining the misfortune of his brothers.

The same is true of our life, we come naked and we leave naked. The only thing you can take to the other world is knowledge, that will lead you back to your father and king. The Creator, Who sent us to Earth to upgrade our Energy-Intelligence. But if we spend our time amassing fortunes more and more, though none of it you can take with you to the tomb. If you think that this world is the only life we live, you are in for a surprise and not a nice one.

This life as we know represents 1% of our real life, and the consequences maybe for our actions that reflects in all 4 dimensions. This opportunity that the Creator gave you, when you were reincarnated. You come to harder times if you degraded your level of Energy-Intelligence or soul. You came back as a lower form of Intelligence and there is a certain level that if you lower beyond, you will not have the minimum level of Energy-Intelligence required to your next reincarnation. You go to the next animals, foul or fish then the next is plants and last minerals.

Beyond that you reach a level of no return, which is the center of the Earth serving as its fuel, toward the Lava of (Guehinom) purgatory where the energy is burned, or if you don't go deep you might find yourself in a gas tank of a car, where again you will be consumed to smoke. Just sit down and think. We know for a fact that fuel comes from fossil of animals and humans and not all humans leave residues of fuels, depending on the purity of his energy.

The return of the Soul or Energy-Intelligence, that comes back is attracted toward a new mass, without form. It is called the clothing of the light. That energy is directing the mass without form, to form according to his past knowledge of human form. This intelligence has its full knowledge and awareness with all the secrets of what he learned after he left his other life. The moment the baby emerges from the womb, at that moment he cries because he comes out to a different environment and pressure, so at that moment his life receives the influence of "The House" (stars, planets, and more) that he is born under.

This is the negative influence on the physical body. We know that the moon affects the tides, so the body is formed mostly of fluids and the moon and other planets will affect the path of life of that person born under that specific influence. To understand see fig. 5 The influence of the planet and the stars.

FIGURE No.5: The Influence of Planets on Humans and all forms of life on EARTH.

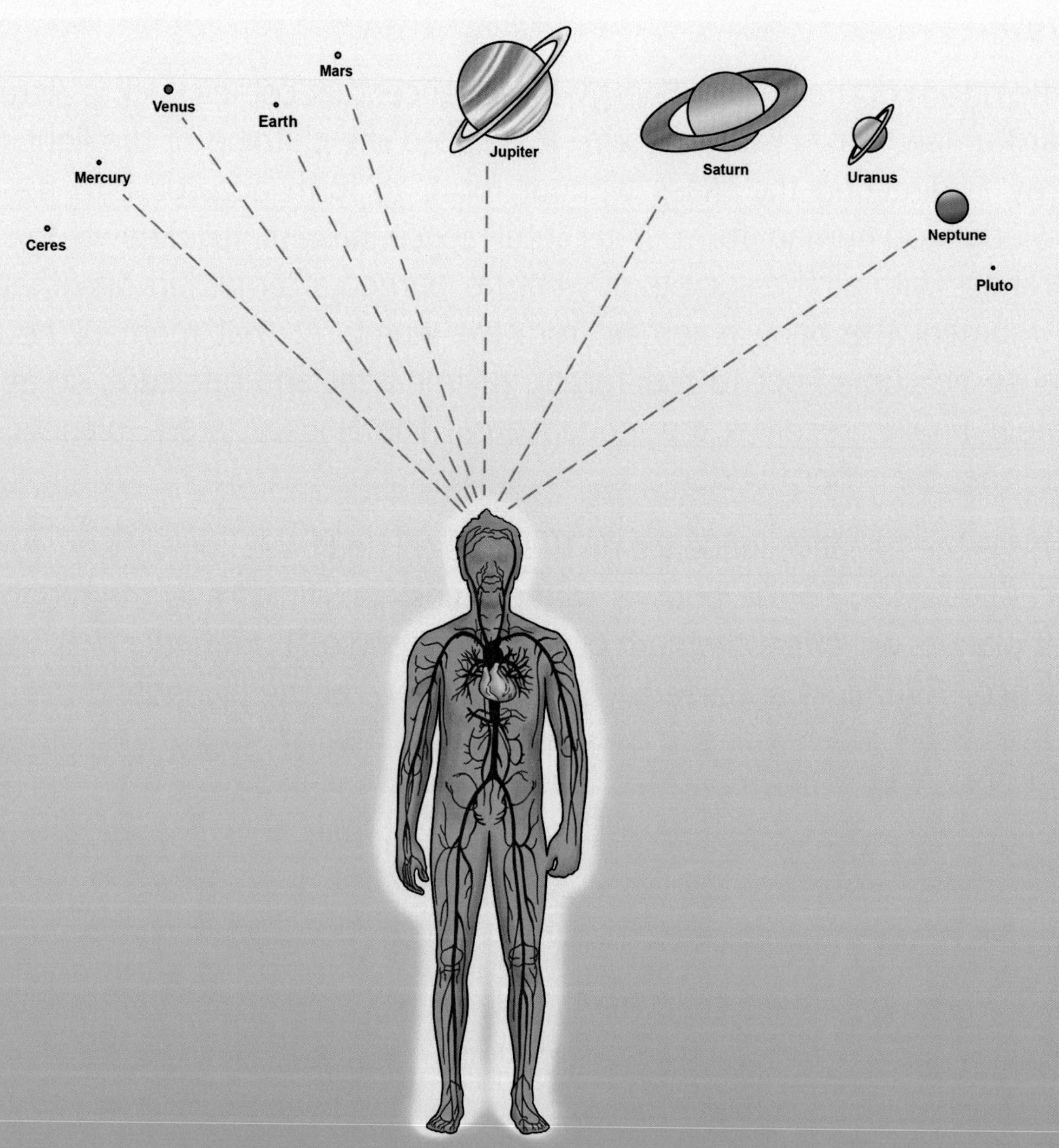

To understand this figure you see the earth and the position of the planet at the same moment, if you are under a heavy influence or house, you will be in good standing with lower pressure and it will be easier for you than other persons born under light influence than when he is under heavier pressure, it will be difficult for him, to reason. The easiest solutions will be acceptable and that will direct this life toward his predestination, of ups and downs. We are attracted toward troubles that are caused in our lives because of predestination toward good and bad. Now there is a way to redirect your predestination. (See the chapter of the Commandments toward a sure path).

"G-d said to Abraham, go away from your land, from your birthplace and from your father's house to the land I will show you." Here G-d is telling Abraham to leave the house of his father, not just the physical house, but the house of the stars he was born under. Which is the spiritual negative influence, and the physical negative influence of idolatry of that country.

(Genesis 14:14), G-d said to Abraham, "Raise your eyes and from the place where you are now (standing) look to the North, to the South, to the East and to the West. For all the land you see, I will give to you and to your offspring, forever." G-d said to Abraham to raise himself, not his physical eyes but spiritual eyes, toward the stars, meaning his Energy Intelligence, leaving his body and to see from high above the influence of the stars.

When a baby is born he is under the pressure of the stars an influence called Klipah (Shell), this is the shell of negativity that is in a sense our body due that the body is compared to a shell and it is negative because it has material physical needs. For example; food, going to the toilet, sleep, dress, to wash, and more... and our life is called the Light of the line because it has a beginning and an end, it is also very important because without it, you cannot incorporate the Endless Light that is represented by the Circle, that has no beginning and no end.

So the Light of the line is like the Antechamber to paradise (the Incorporation of the Endless Light or to purgatory - Guehinom - center of the Earth that is the endless of the night the world of darkness. I'll explain: the light of the line is compared to an athlete that has to jump an obstacle and has to fall in a circle, if he doesn't take enough spring in the straight line before the obstacle, he will not overcome it. He will have to come back to the next Olympics to retry.

The same is our life if we don't try hard enough to overcome the obstacle (Temptations of Life). We cannot reach the target, the circle or the endless light, or light of the circle like the athlete, we fall in the obstacle, we will have to come back to the start of the line, or the beginning of life, the light of the line. So the light of the line is important to upgrade the purity of our Energy Intelligence.

The Kabbalah sees the body divided into 248 parts (limbs) and within those parts, 365 main veins and arteries. There are three different souls or inner Energy Intelligence in a human body. NARAN-נר"ן (NEFESH-נפש, ROUAH-רוח, NESHAMA-נשמה) One is in the blood (that is the negative one) and called in Kabbalistic terms "Sitra Achra-סיתרא אחרא" (the other side) which is a word Aramaic because it is said that "Because the Soul of the flesh is the blood and from it are coming the four bad measures:

1) Ka'as -כעס(anger), Ga'ava-גאוה (pride), that emanate from the element of fire.

2) AAVATH HATAANUGIM –אהבת התענוגים(appetite for pleasures) from the element of water.

3) HOLELUT VELEITZANUT, VE HATIFFEERET VE DEVARIM BETELIM –הוללות וליצנות והתפארת ודברים בטלים(frivolity, scoffing, boasting and idle talk) from the element of air .

4) AATZLUT VE ATZVUT –עצלות ועצבות (sloth and melancholy) from the element of the earth Again all those are influenced by the influence of the house of planets you are born under since the blood is liquid, the planets have influence over it. (see fig. 6+7).

FIGURE No6: An Example of Surrounding Energy Intelligence.

Here we can see two different dimensions of the Human Body,being the Skin&the Blood Vessels. The skin being EMANATION &the Blood vessels,CREATION.

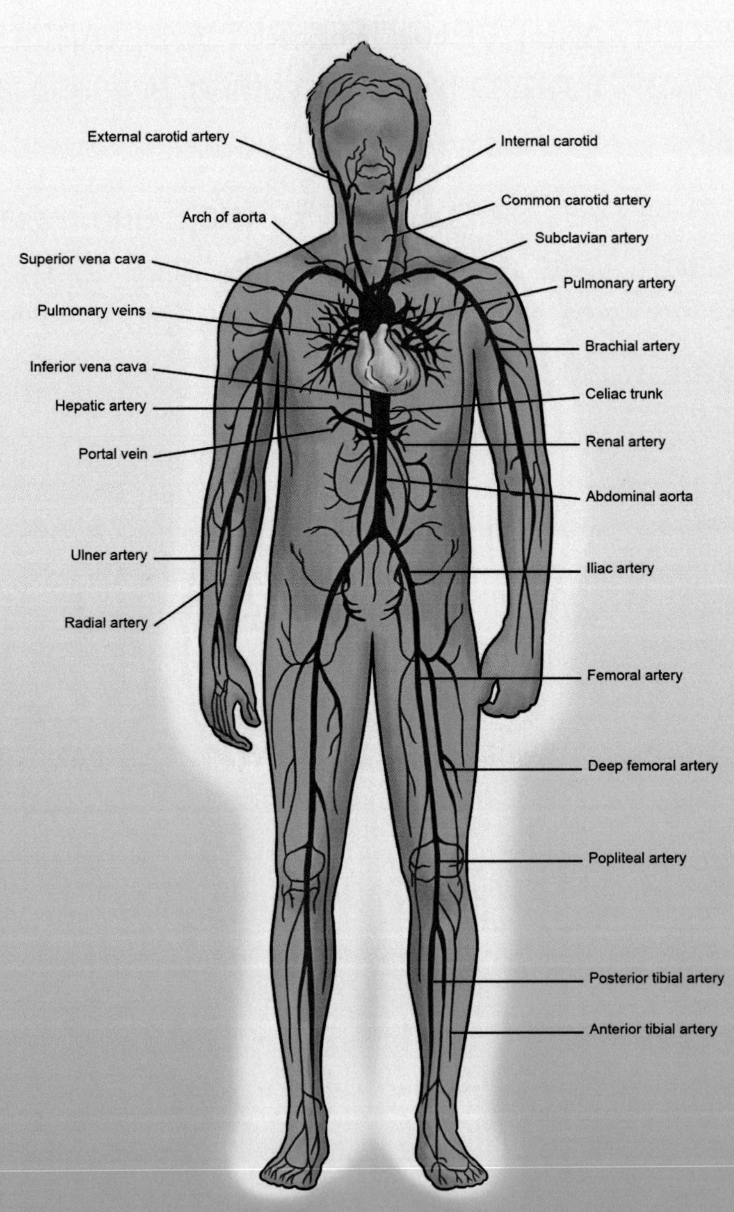

FIGURE No.7: The Bones,Squeleton.

This is the Inner dimension of the Human body,equivalent to the world of DEED,PERFORMANCE.

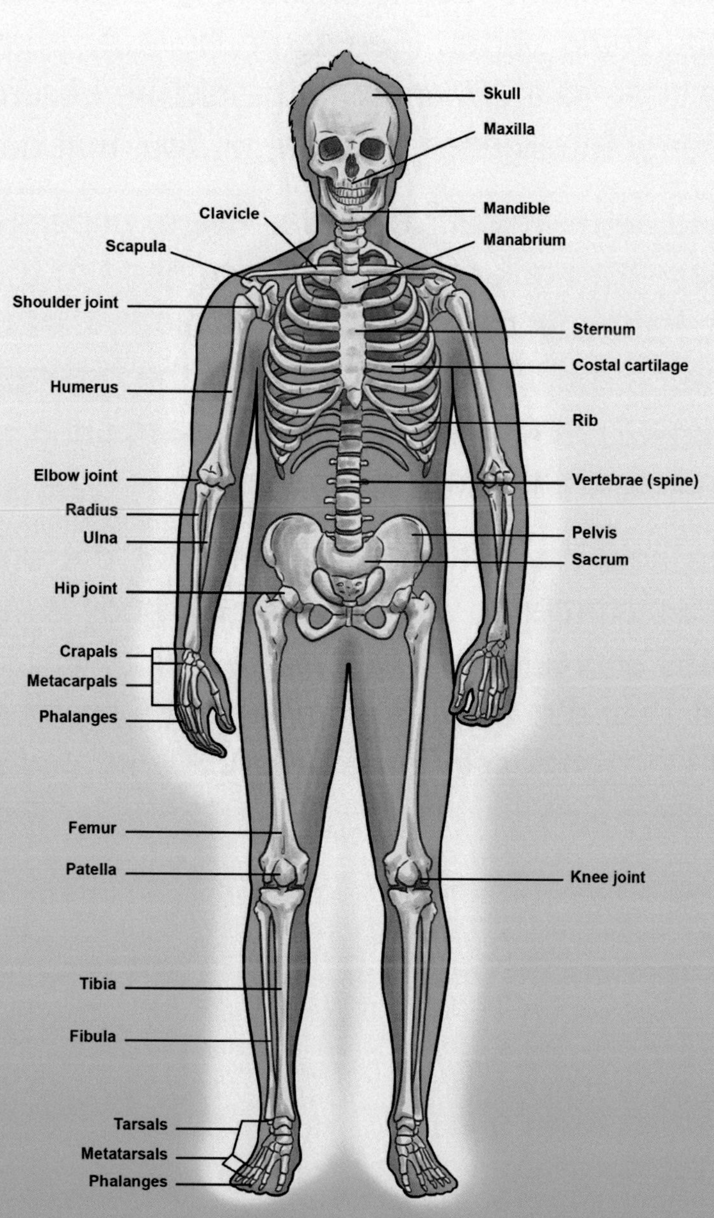

The temple of the second inner soul is the heart (see figure no. 6) and the third is the liver (see figure No. 10) now all those are the inner Energy Intelligence and the surrounding is the one we call the Aura, that surrounds the body and directs the brain. Some scientists call it the mind, but this is only the scientific term, and not the original that emanates from the Creator's essence, to understand better. (See figure No. 10). T'hat is the reason that when we come close to a T.V. antenna the picture becomes better, is to our surrounding Energy Intelligence, or Aura, or Soul that does that effect.

To understand better the universe we have to understand first the human body. The Zohar -זוהר(book of splendor accredited to the first century sage and Kabbalistic Rabbi, Shimon Bar Yohay-רשב"י", that deals with the Light of the Line, and on the 15th century one of the greatest Kabbalistic of our time, the ARI-הארי הקדוש (the lion) Rabbi Isaac Luria –יצחק לוריארבי that explains the Endless Light and the spheres.

They explain how the body is a blueprint of the universe. We tried to understand their statements, we know that the body is made up of an infinite number of atoms and cells, and that every cell is similar to a solar system, we understand that the atom is composed of protons, neutrons, and is surrounded by electrons, (see figure No. 8).

FIGURE No.8: Human Atoms of a Cell.

The size is not correct nor the proportions. The light around the atom is the"surrounding Light"(Electrons). The center is the "Inner Light"(the proton). The rest being particles(Quarks,Baryons,etc.)

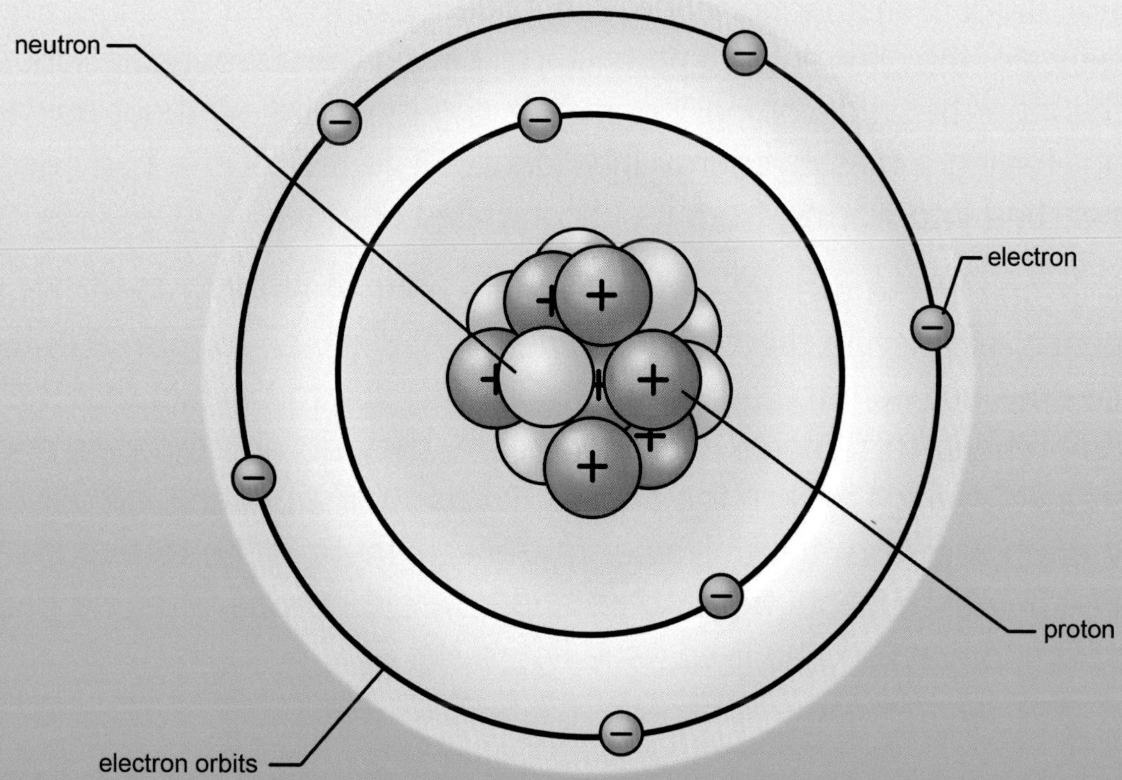

For example, the skin is made up of different cells that of the muscle, and the muscle from the veins form the bones. So then you have four forms of intelligence if you see figure No. 8, it shows the atom with a center and around it a light, that light is called surrounding light of surrounding Energy Intelligence that communicates with the same intelligence as itself, and the same with the veins and the bones and muscles and skin.

Now each of those 248 parts of the body has its own intelligence. As an example see figure No. 10. So when a body dies, the body's soul or surrounding Energy Intelligence leaves the body and, the inner Energy Intelligence stays without control of that higher Energy Intelligence, and then the inner Energy Intelligence of the different 248 parts starts leaving the surrounding light first, then the inner begin decomposing and leaving its atoms, Without guidance, thus not having a master energy intelligence. The rest of the body decay's by converting into worms, from its uncontrolled energies.

This is one of the example how we could reincarnate in a lower form of intelligence, not having the sufficient energy intelligence to form a human . Science does not want to understand it this way, due to the fact that they do not comprehend this Energy Intelligence and they do not know its origins. So instead of trying to reach the planets, let them first understand our body, that is the blueprint of the universe. In Genesis there is the tree of life that explains the descendants from Adam and how old he was when he had his first child and how long he lived, (see figure No. 9),

FIGURE No.9: GENEALOGY

LIFE-SPAN

	Year of Creation	Lived	Until	Fathered
1.- ADAM	1	930 years	930	Seth at 130
2.- SETH	130	912 years	042	Enoch at 105
3.- ENOCH	235	905 years	1140	Kenan at 90
4.- KENAN	325	910 years	1235	Mehalel at 70
5.- MEHALEL	395	895 years	1290	Yered at 65
6.- YERED	460	962 years	1422	Chanoch at 162
7.- CHANOCH	622	365 years	987	Metuselah at 65
8.- METUSELAH	687	969 years	1656	Lemech at 187
9.- LEMECH	874	777 years	1651	Noah at 182
10.- NOAH	1056	950 years	2006	Shem at 500

The reason they gave us all the details, is to my opinion, that when for the first time we ejaculate semen, that particular time will determine the biological clock that starts ticking that same moment, and that will cause the length of our life span. The semen is an extract, mainly from our brain and it is called in Kabbalistic terms MOCHIN-מוחין, MOAH-מח in Hebrew is brain, but it comes also from the Energy of all the cells of a body.

The semen contributes, the veins, bones, and nails and the rest is from the ovum of the mother. When a mother procreates, that same moment what she sees or feels will affect the physical appearance. In the midrash it is said that Jacob-יעקב made a deal with Laban-לבן his father-in-law in Genesis in Parashat Vayetze.-30:33) פרשת ויצא) "In the future this will be a sign of my honesty. I will let you inspect all that I have taken as my pay. Any goat that is not spotted or streaked, or any sheep without dark markings that is in my possession can be considered stolen."

"Agreed," replied Laban, "May your words only come true!" That day, Laban removed the ringed and streaked he-goats and all the spotted and streaked she-goats, every one with a trace of white. He also removed every sheep with dark markings. These he gave to his sons. He then separated himself from Jacob by the distance of three days journey.

Jacob was left tending Laban's remaining sheep. Jacob took wands of fresh storax (plants), almond and plaire. He painted white stripes in them by uncovering the white layer under the wands (bark). He set up the wands that he painted near the watering troughs where the flocks came to drink, facing the animals. It was when they came to drink, that they usually mated.

The animals mated in the presence of the wands and the young they bore they ringed, spotted and streaked." That's what this paragraph wants to explain and not really about goats and sheep, but rather of life's secrets. This explains that the mother is responsible for the physical appearance of the baby.

So it is important for the parents during the moment of conception to have pure thoughts, love, tenderness and to plan the perfect timing. (See Kabbalistic calendar for 200 years [1900 - 21001 available by same publisher). Because during life's period there are negative and positive time zones, so even to plan a baby it is good to plan that he or she will be born in a positive time zone. It is especially important that the wife be clean at least 7 days after her period of menstruation as explained before. It is also important that the husband and wife be in the best condition possible so the embryo will not receive negative energy causing the embryo to receive wrong data and grow deformed, such as Downs Syndrome or other deformities.

The parents are responsible for the misfortune of their future child. Like the parable I mentioned in the first chapter before putting the plug in the socket, read the instructions, especially to conceive a new life it is very important. Get all the instructions you can get which will eliminate the higher percentage of wrong things occurring to your future child, and it is a great decision to give life, and not just for sexual enjoyment. After the baby is born, they should have an environment of love, peace, and understanding.

You should not be nervous close to the child, he/she perceives everything, he/she is similar to a computer without a program. It is good to gently massage the little baby when you are calm and to do it gently you will give him a good sense of security and calm, and his future behavior and health will depend on it. It is important not to over do it, and not to spoil him/her because this will damage him/her because even from too much goodness without taking consideration the negative will have a negative results but like freedom, it is very important.

There must be a limit to our freedom, we have to set a guideline, for example, eating, sex, etc ... Too much sex makes it become monotonous. The more you give it to your body pleasures, the more it craves for new enjoyments causing you to degenerate. It is also the same in politics, too much democracy causes anarchy. The quantity affects the quality. A perfume extract is powerful, the more you dilute it, the less its strength.

The same is true for all life forms, too much water will kill a plant, and so forth. To increase the Klipah-קליפה (Shell), and make it strong, and so weakening your soul or Energy Intelligence, imprisoning it with a heavier shell. As it is true for a car's engine when you put too much oil in it, it will not run properly.

Similarly the body will not run properly in a condition of ill-health. In this state, the brain will not physiologically react in syncopation due to pain. It appears to override the ability to respond. Eventually various means will be undertaken for pain elimination.

At that moment you cannot think straight and that is what happens to people who get addicted to drugs, they block pain or sadness and feel good for a short interval. Then the pain or sadness becomes stronger, then they need more drugs to return to that status, to break away from the pain sadness or other situation. That is called addiction, drugs do not resolve the problem, just postpone it. When you have a tooth decay, it will not go away, you must go to a dentist for treatment, the same thing is true of our lives, when we have a problem, you have to go to the root to resolve that particular problem, physical or spiritual.

If you take good care of your car, it will take you everywhere without trouble; don't treat it well and a breakdown will result. The same goes for your body. The more you deny pleasures, the more you increase your Energy Intelligence, the less the shell (body) is thicker by the negativity that surrounds your inner self energies and makes it possible of connection between the surrounding light and the inner light. In the next chapter, I will explain this subject extensively.

THE BODY

THE SOUL

MIND/ENERGY

INTELLIGENCE

Its power and

Limitations

הגוף הנשמה

כחם והמגבלות

CHAPTER IX

The Body, The Soul (Mind/Energy Intelligence) Its Power and Limitations הגוף הנשמה-כחם והמגבלות

As mentioned in the first chapter, it is speculated that we are using only 5-10% of our brain's capacity. The brain is like a computer. It is many times more powerful than a Cray super-computer, so why don't men use, 50% or more? It is said that Einstein used up to 10% of his brain capacity. Well, the Kabbalah is that program or menu that will allow you to reach to the other 90-95% of your memory bank. Do not forget that your Energy Intelligence is a spark from the Endless Light of the Creator, and we have access to most of the secrets of the world and of the world to come.

Meaning, that when you are in a spirit form and no longer possess a body.

(In Exodus 33: 1 8) Moses said, "Please let me have a vision of you (let me comprehend your unique nature) glory", begged Moses. G-d replied, "I will make all my good pass before you and then reveal the Divine name (Tetagramaton) in your presence. (But still) I will have mercy and show kindness to whomever I desire." G-d then explained, "You cannot have a vision of my presence. A man cannot have a vision of me and still exist (and live, not man, or any other living creature can see me).

"Let's try to analyze this paragraph. Moses is the highest of the prophets and reached one gate of awareness and understanding below the Creator. The only thing he could not see was the Creator and stay alive. Let us try to understand the inner meanings, first Moses ascended to Mount Sinai for 40 days and 40 nights. (See figure No. 11). Which shows that there are ten levels of awareness on every one of the four worlds or dimension, every night Moses ascended (spiritually), his Energy Intelligence will reach one more level of awareness and during the day, he will write the teachings, and on the 39th night, he wanted to see the Creator's essence, to go beyond the world of "Atzilut" to the 40th level in EIN SOF-אין-סוף, because once there, he will comprehend all life's secrets including the most daunting of them all, the Creator Himself

This is the meaning of entering Paradise, he will understand and will not come back to Earth or to his body. It is similar to a person who's in prison, the moment he is free and goes home, he does not want to go back to prison. The same is true when we will feel and understand the Creator, we will not want to live in a world of illusion that we live in. For a better understanding. When we are young kids, we play Daddy and Mummy or the doctor and the patient but we are serious and we believe in it as though it was real. We even cried when things did not go the way we hoped or wanted. To parents, we see it and we know it is only child's play.

The same thing is true of our life. We play the soldier, the bread winner, anger, sadness, etc..., and seen from a higher level of awareness, it will be as we see our children playing. Just an illusion, a game, even though it appears real. As a parent you will not want to play children's games. When you understand what life is about, you will not want to play it again.

The world, when seen from an airplane, there are no frontiers, criminals, police, courts, governments, troubles, hospitals. It is seeing a different world from above, just a few miles higher. But down, back in this world, are all life's troubles, pollution and more .... we only flew in an airplane, not in another world or other dimension. So from another dimension, we see such peace. Why would someone want to come back? For trouble? Pain? Suffering? You want to stay in that level because you do not want to play anymore, "a kid's game." Another comparison. The lower level souls that cannot reach high levels of awareness, it is a punishment to have to come back (to reincarnate). Life is compared in the Midrash-מדרש (The scriptures) like the shadow of a passing bird. So we have to learn how to utilize "The Light of The line", is our passing through this physical world, in order to reach the endless. The Creator gives us the tools to do good or bad, we have a free choice. But He also gives us limitations. (See the chapter of The Commandments).

Let's assume we have $100. We can use it to clothe our children, or use it to buy alcohol or drugs. It is our free choice. I will cite a parable: There was once a very poor merchant who had a wife and ten children ages 1 to 10, all too young to work and help the father. He had a small factory that manufactured candles.

One day a woman came to buy a candle and asked the poor man, how he was doing? His heart felt very heavy and wanted to pour his heart out to the nice lady whom he had never seen before. After hearing his pitiful story, how he couldn't even properly feed his children, the woman opened her purse and took a diamond, and gave it to him saying "I know of an island where there are diamonds like this everywhere you go. You just have to bend down to pick them up. But you have to take a ship that passes through, once every six months and will bring you back six months later."

The poor man thanked her and went to his wife to tell her what happened. After discussing it, she said, "Look, we are not making it anyway, so go try your luck on that island and come back, I'll manage

here with the kids." The poor man took the ship that same week and left for the mysterious island. A month later he arrived at the island. The moment he disembarked, he saw a diamond on the ground, he picked it up and placed it in his pocket, then another

and another, until he filled his pockets, then he emptied the sack that he was holding his clothes in with his food for the trip and filled it with diamonds.

The sack was full and he felt very tired so he decided to sit down and rest. He remembered that he hadn't eaten and was hungry, he found a hiding place for his bag full of diamonds, took one diamond and went to buy a hot meal. He stopped at a food store and told the owner that he wanted to buy some food, if he could pay with a diamond. The owner said, "I am sorry but the diamond has no value here, the island is full of them." But he felt sorry for the poor hungry looking man, and said, "You know what? I like the shape of your diamond, I'll place it in my aquarium." So he gave him some food and left.

He found a tree and sat beside it, night fell and he saw that the houses were made of diamonds too! The only light they had was coming through the reflection of the moon, through the diamond walls of the house. He had an idea to try and sell candles, perhaps it would work. So it was the next morning he went and got wax and made candles and went everywhere to sell them. His luck turned and the idea caught on, they had never seen candles before. Our poor man became very famous and rich.

When the time came for him to return to his country, the ship arrived and was leaving the next morning. So all the inhabitants of the island made a party and all his money was invested in candles, so the villagers filled his baggage with candles. He embarked the next day and forgot completely aboutthe bag with the diamonds that he had hidden in the cave.

Finally he arrived home with his wife and kids welcoming him with hugs and kisses. When his wife opened his suitcase she found only candies and began crying saying what are we going to do with candies, we can barely sell the ones we have. Where are the diamonds that you went for?

The same thing is true when we are in an Energy Intelligence or soul form. We want to come to this life, to correct and acquire more knowledge in order to reach the Endless Light, but the only way to have another chance and upgrade is to reincarnate and correct the soul. TIKUN -תיקון(Karma) than we come to earth for the diamonds. The TORAH-תורה (Bible) and the Kabbalah-קבלה, but instead we are busy with the candles (material things, money, cars, homes) we need them but to a certain measure only and not to spend all our trip (life) only amassing candles.

The knowledge of the Torah are the guidelines for the trip and the diamonds are the TORAH and the Kabbalah, that will let you elevate yourself to a higher level of awareness and purity, the candles illuminate very poorly. The Torah knowledge is the only thing you can take with you. So if you apply this moral lesson to your actual life, you will improve and will reach a higher degree of intelligence and awareness. The main rule to attain that knowledge is that you have to eliminate four things: a) Anger, b) Sadness, c) Pride, d)Selfishness.

I'll explain why!!! All the above are curtains of negativity that will block the possibility to ascend to a higher level of awareness. Let's assume you are angry or sad, everything looks dark to you, your energy at that moment becomes perturbed energy. (See figure No. 10).

FIGURE No.10: Example of Perturbed Energy.It occurs to an angry person,depressed,sad and more. The black dots and black lines shows absence of Positive Energy(surrounding Light) causing voids of protection from Negative Energy,in certain cases;Evil Eye.

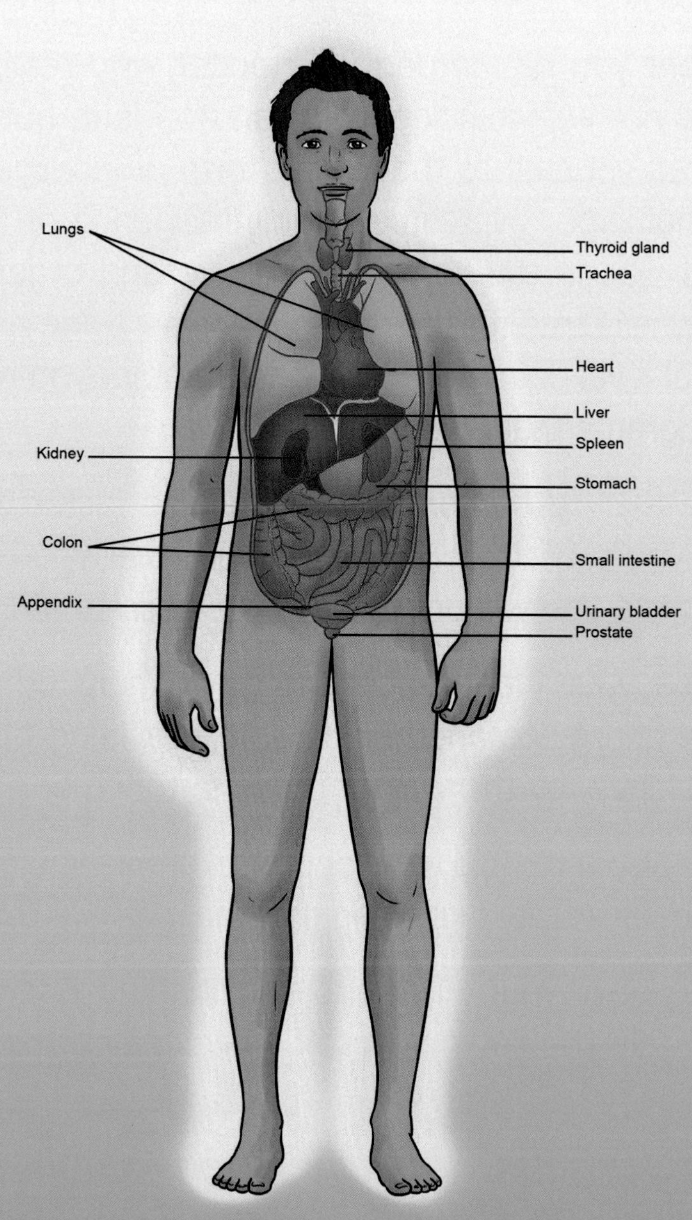

It will be like short-circuiting a computer. On the screen, thousands of characters will be shown, but the computer will not function normally. The same is true of our intelligence energy, the moment we are short-circuited. We do not see straight, we will react irrationally, in a moment of anger you took a steel pipe that was near by and hit the person with whom you were angry, and killed him.

At that moment you discharged your anger, you realized, my G-d, what did I do? But it is not just coincidence or an accident, but in that moment of anger you lowered yourself to the lowest form of intelligence. The mineral's intelligence of that pipe, overcame your intelligence of that pipe, felt your vulnerability and was in the same radio wave and ordered you to hit that individual, the same is true for the other; sadness, pride and selfishness, they block your normal level of intelligence and do not let you reach a higher level, due to those programs that are harmful.

Beyond these four restrictions you could rise faster to a higher level of awareness, and you can attain amazing levels, that you have never seen life from that angle, and you will understand the secrets of life.

a) How to use your intelligence energy to help others, you can even transmit positive energy to another person, providing him with a cure that pills could not achieve, by the way of prayers, touch and so on.

b) The mind is also a terrible weapon, for example you could enter anyone's mind and insert ideas, without knowing that the idea was implanted in his brain.

c) You could control your inner and surrounding Energy Intelligence and dematerialize or tele-transport yourself. To explain further, since the body is made up of atoms and every part has its inner and surrounding light or intelligence energy, (See Figure 6, 7 and 10).

When you reach a high level of awareness you have to become master of your intelligence energies, in order to control all the energies, order them to leave your body form, by doing so, leaving blank atoms, liken to the ones of the air, because you don't leave a vacuum space.

So only the energy leaves the particular atoms, and since the air is atoms and conductors, you can tele-transport your energy intelligence to another part of the world, or the galaxy, instantly on new blank atoms but with the same Energy Intelligence.

The idea of tele-transportation that we see in Star-Trek has been taken from the Kabbalah. You do not need a transporter, your brain can reach this level of intelligence, and know and control every part of your body in the subatomic world.

To understand the world of miracles, and the cosmic forces available to us, we have to eliminate these four conditions and restrict the body of its pleasures to a maximum.

Moses was the greatest prophet of all time. He started when he was in exile, and saw the burning bush, and the presence of G-d, the Creator, and from that moment on, he began to ascend to a higher level of intelligence. A lot of secrets were divulged to him by the Creator Himself. He learned the powers of the universe, but without understanding fully how this actually could happen. Only through using the sacred name of G-d, the(Tetagramaton).

There are ways to do the miracles through secret formulas, hidden in the Kabbalah, but when he ascended to Mount Sinai he learned all the secrets of the universe but one, G-d Himself. Above the world of ATZILUT-אצילות (Emanation) the unknowable G-d cannot be defined. As written by Maimonides: "He is the knowledge and knower" and it is not within the power of any man to comprehend clearly, as it is written "my thoughts are not your thoughts" we know only that we are made from his essence (Energy-Intelligence in a pure form), and everything in this Earth is filled with His Glory, so nothing can existor be without His essence.

This Energy-Intelligence is only a spark of the Creator, and there is no way to see it or measure it, so lets content ourselves to derive from his Energy Intelligence, and try our best to dedicate ourselves to draw as much as we can or are able from Him. It is also known, that there is no way to hide from the Creator and everything we do or know or think, He instantly knows. To help explain this further here is and example: Computers with a modem can be connected to a super-computer but there are a lot of access codes that you will not be able to break, but at the same time, the super-computer can draw from you, information from your data bank instantly because you are connected to it by the same way it is connected to you, the Modem! To the Creator, we are connected by His essence that is through the breath of life mentioned in Genesis 2:5 and breathed into our nostrils a breath of life.

Moses was a very humble man, down to earth. He never cared for himself all his life since he left Egypt. He dedicated himself to study and to free the nation of Israel. Many persons have said that Moses acquired his powers from the Egyptians. It is a ridiculous theory, because he was a prince warrior in Egypt and not a magician. Only the magicians used sorcery, and the Egyptians were using the one column system, the left (the negative). The Kabbalah is a three column system, and his powers were above any magicians in Egypt. (See Exodus).

The knowledge he acquired from the Creator. He learned the dates of cosmic events, today's Jewish Holidays and used masterfully those cosmic energies inspired by the All Mighty. The Creator's source of Intelligence, to direct those destructive powers in Egypt and caused the ten plagues. The stars and the planets in the cosmos are called the army of G-d, the angels.

The Angels are in many forms, for instance; letters of the Hebrew alphabet are also angels. It is stated in Genesis that G-d said, "Let there be light, and there was light" meaning his words or letters caused the deed. SO it is said in the Kabbalah thatthe letters are angels, and to every action there is a reaction and every action starts with a thought. The thought itself is an action, due to the fact that there is nothing you do before you think about it. That is the reason, it is important to have a clean and liberated mind, and not to think about vengeance, hate, greed, it will lead only to unpleasant actions.

Going back to the angels, the stars are angels, the mighty army of the Creator. Because they exercise influence on life of the earth, on the negative side,the ones that influence us first at and immediately after birth. Also angels are intelligence energies, detached from the unity of the EIN-SOF, אין-סוף the Creators brilliance. They are in a pure form of energy. Now an energy intelligence of an angel can be in contact with the human mind or energy intelligence and he will show himself depending on the individuals perception of what an angel looks like. If there are two people, he can show himself to one, and to the other he will stay invisible.

People that can see-in day light and awake, people who have passed away long ago, he will see that person even dressed in a certain way. But only that person will see him, because his soul or Energy

Intelligence contacts the living person in a wave that only he perceives, and remembers even his way of dressing. He is not imagining, he is actually being contacted by that source of energy. So we perceive angels in human likeness with wings.

Moses never took credit for any of the miracles, which the Creator ordered him to do. He was a very humble person, good hearted, and he cared for the Israelites as though they were his own children and family. He is considered the greatest of the prophets. The Creator never let him see the promised land because of two mistakes that he made.

Let's analyze this. Why was the Creator so harsh on him? comparing this with an example: You have a brand new white shirt, suddenly you dirty it with a single spot. You automatically change the shirt for another one, because in a white shirt it is very visible. But if you put on a brown or black shirt, the spot will not be noticeable. So was Moses, he was so white in purity that those two tiny spots were noticeable. The Creator had the highest regard for Moses due to his purity and he washed his spots immediately after taking him. He was higher in purity than an angel.

In those times, G-d did many miracles for the children of Israel, and forgave them many times, so why not now? We are punished almost immediately. The reason is explained with an example:

When we are born, and are babies, even if we break something our parents will not punish us because we are too young to comprehend. But when we grow up, if we make the same mistake, we will be punished. Israel was in its infancy at that time, first time a nation, so G-d pampered them. But they were 40 years in the desert, meaning that at 40 years a man is then wise.

That is one of the meanings of the 40 years in the desert. 40 is the age of wisdom. Another example of the inner means in the Kabbalah is shown by Ari Hakadosh (Isaac Luria). Our ancester Jacob and the 70 members of his family descending to Egypt, is in one of the 70 facets of the bible the act of mating, the 70 seeds of fertile sperm going down to Egypt(the woumb).

The ten plagues are the 10 different periods of pregnancy and the birth pangs every 10 minutes, before birth, when the waters of the woman's womb break, just before giving birth. Those are "the waters of the Red Sea drying", they walked on dry land is when the baby emerges out of the mother's womb to a dry world, and the MARAH -מרה(the bitter source) that's the children of Israel when they cried for water, the baby also cries when he comes out into the air because of the bitter pressure of the influence of the stars, (after the Marah) the 8 days of Passover-פסח are the eight days that we wait for the circumcision, and the 3 days to sanctify oneself before Shavouot, are the 3 days that the baby is in danger after the circumcision and we have to take care of him The 40 years that the Jews stayed in the desert, has another additional meaning in that the baby cannot see for 40 days after birth. Also the Succoth Holiday means, the Marriage of Israel with G-d like in a wedding we are covered with a Tallit (prayer shawl) with four corners over the heads of the newly wed called "chuppah-חפה". Here we are covered with the succah.

King Salomon in the song of songs compared Israel to the bride and G-d as the groom. The 7 days of the Succoth-סוכות equals to the days that a Jewish wedding is supposed to last(sheva brachot-שבע ברכות).

Also when a person dies, the family sits Shivaa,(seven days of total mourning sitting on the floor) here we commemorate the reincorporation of the Soul with G-d's essence, or the Soul marries the All Mighty's Shechinah-שכינה (Holy Spirit). Here we can understand that every thing in life has its parallele from a single atom to the whole Universe and finally to the One and Only Mighty CREATOR

These are only a few of the 70 meanings that our elders have said that the Bible has. And all of them are true indeed. Another meaning: ABRAHAM & SARAH matter and antimatter. Today scientists are trying to create Anti-Matter,it is also said that it is very costly and that it will take hundreds of years to make an insignificante amount. The secrets lies in-between Abraham & Sarah. .The clue is that Abraham-אברהם was called Abram-אברם and Sarah-שרה,Sarai-שרי. In hebrew-י yud equals 10. G-d said to Abram "from now on you will be called Abraham and Sarai Sarah"it mean taking five from the yud of Sarai and add it to abraham and the other five that equals Hay-ה will be added to become Sarah. So taking five from the Anti-matter and adding it to the matter or ½ spin will cause a chain reaction of Anti-matter,thus disrupting the order of: that for each matter there is one Anti-matter. Of course there are still many more hidden meanings that meet the eye and the beauty of it is that all are true. If science turned toward the original scriptures,they will find all universe's secrects.

ABRAHAM, ISAAC, JACOB –אברהם יצחק ויעקב Three column system, 12 sons of Jacob are the 12 zodiacal signs and 12 months of the year. ABRAHAM-אברהם, YITZHAK-יצחק, YAACOV-יעקב,YOSSEF-יוסף MOSHE-משה, AHARON-אהרון, YEHOSHUA-יהושוע being of the 7 lower spheres or Sephirot, ADAM-אדם, NOAH-נח, SHEM-שם, being the 3 higher spheres (In Genesis) "and Moses found grace in G-d's eyes" meaning G-d recognizes Noah's soul in Moses. Noah in Hebrew is the letter (noun)-נ and (Chet)-ח Inverting the Hebrew letters would make Chen-חן meaning charm or grace to find grace in G-d's eyes, his sons are the 3 column system, SHEM-שם being the center, Yefeth-יפת the positive and Ham-חם the negative.

The 70 nations of the world descend from the three sons of Noah. In (Genesis 3:1 1) G-d said, "Man has now become like one of us in knowing good and evil. Now he must be prevented from putting forth his hand and also taking from the tree of life. He can eat and live forever." G-d banished (man) from the Garden of Eden to work the ground from which he was taken. He drove away the man - here we see another example that it is another kind of Eden, meaning in the womb of his/her mother, and the moment he is driven out he breathes the Air of this Earth and come under the influence of the planets, so he cries, and from now on, he has to work the ground from which he was taken. Meaning he has to cry to get his mother's breasts for milk (the ground which he was taken) and from now on he will cry, have toothaches, and all kinds of illnesses, and not the peace of mind that he had in his mother's womb. "And stationed the Cherubim at the East of Eden along with the revolving blade to guard the path of the tree of life." (Genesis 3:1 1) G-d orders the planets that are revolving around the sun and at the same time around the Earth(the Zodiacal signs) to create the pressure of influence that he is under, SeeFig22) the moment he leaves his mother's womb. That negative influence impedes us to draw from the tree of life or the infinite wisdom "HOCHMA ILAAH-חכמה עילאה," that will permit us to understand the universe. In order to do so, one has to overcome this influence that leads us to predestinations and we have the Commandments that will be the guidelines to overcome it, toward a safe haven.

That moment the Israelites forgot, in that instant that G-d had performed a great miracle of the opening of the Red Sea. The child too, forgets his past life experience. This is only one of the many other meanings, reading the whole text, there are secrets of life, that science is still far from reaching.

To fully understand life, you need to realize that it is a life time of research. Every word, letter, vowel or punctuation, has many hidden secrets. For that reason the translation reduces the real explanation to the actual history of our ancestors, without revealing the hidden secrets reducing it to a simple tale in our history.

The main reason I mentioned all of these quotes from the Bible is to show you, that there is more than meets the eye, the research is breathtaking after discovering a new meaning that has not been disclosed by other commentators. Because they didn't find it, or for some powerful reasons they decided not to disclose it. Of course there are many more meanings then I disclose, because this is only a presentation to Kabbalah. In my manuscript, "Sapphire & Diamond" I have written of the esoteric meanings and formulas. The book is written in Hebrew and I will make it my task to translate it after the publication of this book, with G-d's will

THE COMMANDMENTS
And their
PURPOSE

עשרת הדברות,
תרייג מצוות
וצרכהן

Chapter X

THE COMMANDMENTS And THEIR PURPOSE הדברות מצוות וצרכהן

Moses-משה רבינו received in Mount Sinai the ten (10) Commandments in Two (2) tablets of stone. (See figure 12.) This shows that the writing went through the tablets from one side to the other. Why does the Midrash go to the extent to explain this detail? The explanation is, that the width of the tablets denotes the length of time, and we should keep the Commandments in every generation without modifying them even a bit, and I'll explain why.

FIGURE No.11: 40 DAYS & 40 NIGHTS OF MOSES AT MOUNT SINAI.

OR EIN-SOF		ENDLESS LIGHT		
KETER (crown)	1	40th day	ATSILOUT	
CHOCHMAH (knowledge)	2	39th "	EMANATION	
BINAH (wisdom)	3	38th "		
TIF-ERET (splendor)	4	37th "		
CHESED (grace)	5	36th "		
GVOURAH (might)	6	35th "		
YESSOD (foundation)	7	34th "		
NETSSACH (eternity)	8	33rd "		
HOD (glory)	9	32nd "		
MALCHUT (kingdom)	10	31st "		
	1	30th "		
	2	29th "	BRIAH	CREATION
	3	28th "		
	4	27th "		
	5	26th "		
	6	25th "		
	7	24th "		
	8	23rd "		
	9	22nd "		
	10	21st "		
	1	20th "		
	2	19th "	YETZIRAH	FORMATION
	3	18th "		
	4	17th "		
	5	16th "		
	6	15th "		
	7	14th "		
	8	13th "		
	9	12th "		
	10	11th "		
	1	10th "		
	2	9th "		
	3	8th "	ASSIAH	DEED
	4	7th "		
	5	6th "		
	6	5th "		
	7	4th "		
	8	3rd "		
	9	2nd "		
	10	1st day		

FIGURE N.12: THE TWO TABLETS OF THE TEN COMMANDMENTS.

The width of the tablets denotes

That we should keep the commandments during all the length of times.

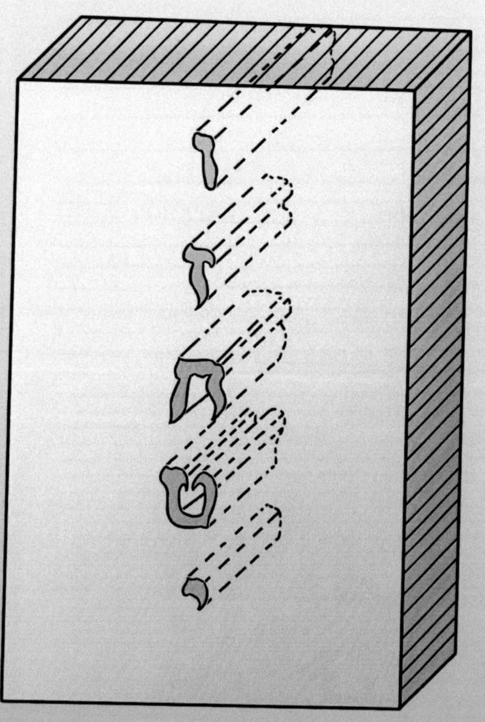

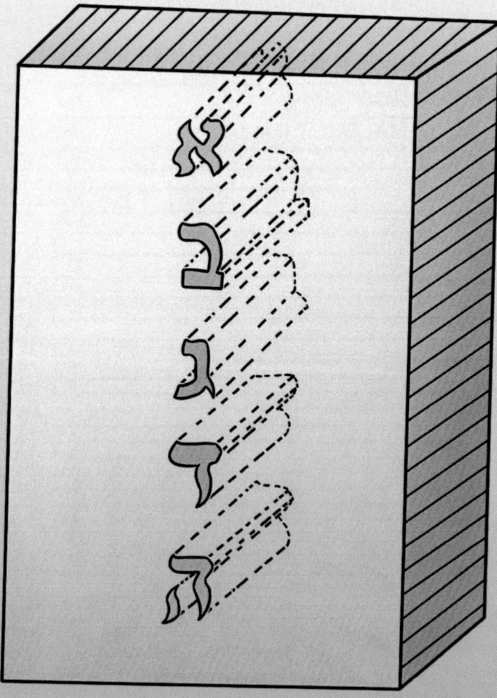

From the 10 Commandments, we derive the 613 Commandments, and they are mentioned in the five books of Moses, and especially Leviticus.

The eleventh century sage and philosopher Maimonides (Rabbi Moshe Ben Maimon-הרמב"ם) the Commandments into 613. From them 248 are positive Commandments (you shall do), and 365 are

Negative Commandments (you shall not do) totalling 613 Commandments. Adding 6+1+3 makes 1 0 Commandments. 9(his is just a small remark)

I already mentioned in the First Chapter, that we are influenced at birth, by the stars and the planets causing our lives to be predestined. G-d said that Israel has not "MAZALOT-מזלות luck to go by, therefore predestination. ARE WE COMPOSED OF

DIFFERENT KIND OF ANATOMY? ONE THAT DOES NOT INFLUENCE OUR BODY, NEGATIVELY? No predestination?

No, we are as human as anyone. But G-d said if you follow my Commandments you will not depend on predestination. Meaning that G-d created also the serpent or temptation, "the star influence". Then we are under a heavy influence, we take a bad suggestion, as being a good one, and it causes us to make these mistakes, which leads us to our predestination. (See Figure No. 13.)

FIGURE No.13: Path of Life & the Influence of Stars over your Destiny.

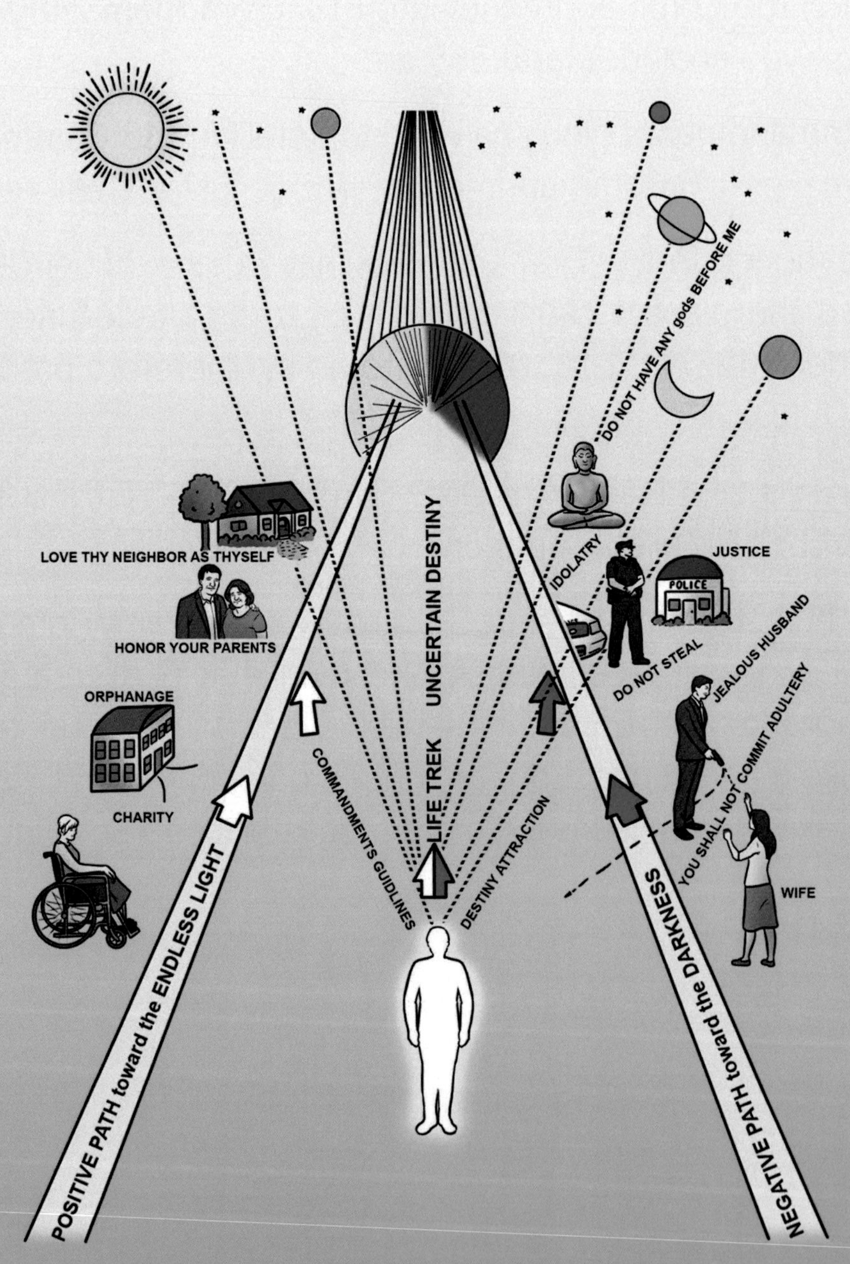

The figure 13 gives you a good idea of how life works, reserving you temptations along your passage. Not everyone knows how to travel through his life life's voyage through happy, sad, good, or bad moments. Behind every good, there is an evil waiting to make you fall. And behind every evil, there is a light on the horizon shining for you, and it is yours to grasp.

The main reason that the Creator, Blessed be He, gave us the Commandments, guidelines to lead us for a life toward a happy ending, toward Eden or re-incorporate the "Ein Sof'-אין-סוף, the Creator. One hour of that final Garden of Eden is worth all our lifetime pleasures.

Lets us analyze some of the Commandments: I know that 613 Commandments sounds enormous. Many of them cannot be performed due to the destruction of the Temple. But the basic one is "Love thy neighbour as thyself." If you perform this one only to the full extent, you accomplish almost 50% of the Commandments. You would ask how? I will proceed to explain.

Let's assume you have a beautiful wife, and love her very much, you do not like others to look at her. There are other people who feel the same. Therefore don't look at other women yourself. If you don't like other people to laugh at you, don't laugh at others. If you don't like other people to come to your home and steal from you, you should not steal.

Do not covet your neighbours property, etc.... this means that, what you do not like for yourself, do not

Do to others. I could go on and on .Dealing with hate.

Let's assume someone you did a lot of good for, in turn paid you with evil. Do not fight with him or her, but rather continue being the same, courteous, and polite. Do not lower yourself to his standards; you are not at his level. In a low level, you are not equal to him he will win. When a person is like that, and he had spoken bad about you. A friend comes and tells you about it, you should answer that you are sorry about how he feels about me, I still think, that he is a good person.

In time your friend goes back and tells him your answer, he will feel bad; you will hurt him with good words without lowering yourself to his level. Hate is very negative, if you wish him bad, only good wills come to him. But, when you do not wish him good or bad, you leave it to the forces of "MIDA KENEGED MIDA-מידה כנגד מידה (Measure against measure) meaning to every action there is a reaction in the metaphysical world, which surrounds us. Thus anything he does will come back against him, and do believe, this will happen. Hate fills you up with perturbed energy that will lower you to a world that you are not good at, and will Influence you toward a bad decision, but on the other hand, if you leave it to measure against measure" to react for you.

If someone stole from you, you go to the police to complain, and should not go to his home to steal from him, because you are not a professional

Thief, and of course you would get caught. (Remember that it is his domain). So, you go to the police (measure against measure) to complain.

Every step in the physical world creates a chain reaction, on the other three dimensions, the metaphysical world. Another good reason to be a good mannered person is when you pass by a person, you salute him, if he salutes you in return, and you always have a smile or everyone. You always to try to be helpful to everyone, poor and unfortunate children, especially to your elders, handicapped, old people in hospitals, who have no families to visit them "Love thy neighbour as thyself').

If you were, G-d forbid, alone in this world, would you enjoy a person with a good soul to come and visit, chat and bring you some help? Yes, you would. If you see an old man carrying a heavy load, help him. Also help orphans; it is a great MITZVA-מצוה (good deed). They are at the beginning of their voyage in this hard world, without a shoulder to cry, when they have a toothache, or if they want a toy like your child has, or to have someone to care for him and call him Uncle, from time to time take him to a movie, to the circus or bring him a gift. Those are innocent children. Who need to feel a little love, a good heart that will give them a family, to be a

Child of G-d. In Israel, they name an orphan," Israel", because he/she is one of his children. Let me tell you what happens, when you are that type

Of person. Everyone you pass by, you will receive smiles from old people, neighbours and children.

Rabbi Judah-רבי יהודה said: Great is the charity, it brings the redemption nearer, as it says "Thus saith the Lord keep ye judgment and

righteousness charity (TZEDAKAH-צדקה), for my salvation is near to on my righteousness to be revealed (ISA. 56: 1) he also used to say:Ten strong things have been created in the world. The rock is hard, but the iron cleaves it. The iron is hard, but the fire softens it. The fire is hard, but the water quenches it. The water is strong, but clouds bear it. The clouds are strong, but the wind scatters it. The wind is strong, but the body bears it. The body is strong, but fright crushes it. Fright is strong, but the wine banishes it. Wine is strong, but sleep works it off. Death is stronger than all, but charity saves from death, as it is written, "Righteousness (charity-TZEDAKAH) delivereth from death" (Prov. 10:2).

So, there is no more to say how charity is great, especially how good the feeling is that you derive from it, and this feeling purifies your soul or Energy-Intelligence, cleans bad thoughts, heightens your self esteem. Let's say a good man goes by, and people refer to you as a good human being, especially you should avoid talking bad about

People, never criticize another who is not present, when one does criticize others in your presence, just stand up and mention that you have an appointment and leave, at every occasion, you do the same, until they understand why you leave every time they slander someone. You can be sure they will never believe anyone that you have made a bad remark about them, they will defend you, and might learn from you. Now, they all have the Mida Keneged Mida-מידה כנגד מידה (Measure against measure) too.

All those eyes smiling at you, wishing you well and respect. It is also an action, and of course there is a reaction, and all those translate to positive energies sent toward you, increasing a stronger shield against the evil eye or bad wishes. Evil eye does exist, but not in a superstitious way, but in a logical one; evil eye is negative, like radio waves, that are transmitted by a certain individual toward you. Sometimes, it is done without malice, because that person, does not know how to control his negative (impulses)waves. When he says something that sounds good, it could turn to bad. These negative wishes or energies, are directed toward your surrounding light, that is your shield.

The positive, is constantly surrounded by negative fields, that tries to penetrate his shield. If the shield is good, no interruptions within it or voids, (see figure 10), evil would not penetrate. If there are voids in the shield, it will let negative transmissions, to penetrate. Compared to radio transmissions, it will override any other transmissions from other stations, if his transmissions are powerful. When it is weaker, your radio will have interferences from other stations. We function the same way. If you are strong minded, not everyone can influence you. If you are not, any person who comes with a bad suggestion, it will seem to you like a good one, and eventually will get you in trouble.

When a good person passes away from this world, people will come to render to him a, last respect, the talk will be: "what a great man he was," he was a TZADIK-צדיק (righteous), a saint and people normally have the tendency to exaggerate. If good, they exaggerate the good. If bad, they exaggerate the bad. This is the final push to a higher level toward the EIN SOF- אין-סוף-Endless Light), or toward Sheol -שאל(Purgatory) on the other side. All have a good feeling.

There is a positive energy in remembering you, that is directed toward your soul, increasing its energy level, rising your soul to a higher level of awareness. For example, a person who is frightened, will see scary things around him, his fear will connect him with the lower realm of intelligence. The same is love. The more you think about a person, the more you love her, the more you want to be with her.

This feeling is an energy that grows in you, and is called feelings, they affect your well being one way or the other. The same is true of the

feelings of other people to your memory when they speak of how good you were. They imagine you smiling and that feeling is a positive energy that elevates your soul, reinforcing your energy level.

All this is "Love Thy Neighbor as Thyself ' a positive commandment. Counteracting a majority of the 365 negative Commandments, being one of the most important of the world of ASSIAH-עשיה (world of-deed). The most important ones are:

EXODUS20:1

1) I am the Lord your G-d.

2) Do not have any other gods before me. Do not represent such, by any carved statues or pictures of anything in the heaven above, on the earth below or in the waters below the land, do not bow down to such gods or worship them.

3) Do not take the name of the Lord your G-d in vain. These three commandments are from a high realm in the metaphysical and physical world, and are the most important. –These will enable you to reincorporate the "Ein-Sof-אין סוף", The Mighty Creator, when you'll be free of this world, to Eden.

The ten (10) spheres, or levels of awareness, are also divided in three higher and seven (7) lower spheres, they are in the form of three (3) for the head, and seven (7) for -the body (The Light of the Line). The energy levels are sub-divided in the same way to infinity.

The commandments:

"I am the Lord your G-d," is a statement that He is the only G-d, He fills the universe with His glory. There is nothing that comes into existence, without His essence or derivations of His essence. When life forms, it is pure. The degradation, we cause it. This is the reason we have to respect every creation.

Do not take food for granted, bless it before you eat, and after you eat. Even food has the Creator's essence. As you know, everything is made of cells.atoms and thus Energy Intelligence. The reasons for the blessings, are not so much that the Creator needs you to thank Him every time you eat, but you do not take it for granted.

The most important reason for saying a blessing, is that since the souls are reincarnating into other lower forms of intelligence, as animals, plants and minerals. We, by eating and blessing what we eat we give a TIKUN-תיקון(correction)of the soul, when it mixes to our energies because all food are energies. So we allow them to upgrade themselves to a human form, a higher one,in other words us. When we pray in the morning and afternoon we say "VIDOOY-וידוי" meaning that with our right fist we hit our heart reciting all errors and misdeeds that we might have done in all the 22 letters of the alphabet. But we recite all these in plural.

So we ask why in plural? Why all these sins that I haven't done? The answer is that we have within us many more reincarnations, that are incorporated in us in many forms. So since "we", the man that is praying also prays for these souls within himself, just in case they have transgressed one of the 22 sins. We can see in the bible it says "Koulanou arevim ze le ze-כולנו ערבים זה לזה" we all are responsible one for the other. Here too we can appreciate the commandment "Love thy neighbor as thyself"

The blessing after the meal (BIRKAT HAMAZON-ברכת המזון) is to thank the Creator. At the same time when you recite the blessing that takes three minutes, you think twice before eating every moment of the day, so it keeps you from overeating and becoming fat. In the same way, you attract illnesses that will cause your brain not to function properly, or serve your Creator the proper way, which directs your life, toward a good path.

Commandment No. 2: "Do not have any gods before me. Do not represent such gods by any carved statues,picture or anything in the heaven above,on the Earth below or in the waters below the land.

Do not bow to such gods or worship them". This is a very important Commandment, I explained this matter at the beguinning of the 1st chapter. I cited the four forms of Intelligence.

The highest one being the HUMAN

The second The ANIMALS,BIRDS etc..

The third The PLANTS,fruits etc...

The fourth The MINERALS.

The last three are created from energies that in some instances might have been human souls or Energy-Intelligence that,When alive reduced their energy to a lower level than 51% positive.

As I explained before,once they passed away,they do not have the Energy Intelligence to elevate themselves below earth's surface, and had to take the lower intelligence form So, people who worship animals, are bowing to a lower form of intelligence, people who worship images, statues, they are lowering themselves to the lowest level of Energy-Intelligence, the minerals.

This behavior, is beyond comprehension coming from an intelligent person. We cannot give a form, or quantify the Creator's qualities, limit His glory, or His powers, or confine Him to a place, when He is everywhere, "He was, He is and He will be. He has no beginning and no end," He is within us and within every living creature. Nothing exists without Him.

The Bible (Torah) mentions His patience, and says "G-d's hand", this is only figurative for a person to comprehend. As Maimonides mentioned in his book,

"Guide to the Perplexed," to give qualities to the Creator is denoting Him with human limitations, and He has no limitations, and it is impossible for a human mind to comprehend. As it is said in the Bible, "My thoughts are not your thoughts."

The Third Commandment:

Do, not take the name of the Lord your G-d in vain." In Hebrew when we pray or read the Bible, the name of G-d (Tetagramathon) is written, but we do not pronounce it. We read Adonai-אדני (Master), even so, we pronounce it in prayer only. We say Hashem-השם (the Name), when we are not praying. So sacred is the name that our lips are not clean enough to utter it. During the time of the Temple, the high priest was the only one to pronounce the sacred name of G-d "Shem Hameforash-השם המפורש" or (the name disclosed) the "Tetagramathon." Moses understood its power and the name was engraved on his staff, with which he performed all the plagues in Egypt, the parting of the sea and all the incredible miracles. This gives an understanding. As an example: If a person insults someone, just his words create a negative reaction, or if you tell someone how beautiful she is, and how you love her, this creates a chain reaction.

Words can change your whole life. Bad words could end in a fight and/ or a death and put you in prison for the rest of your life. With loving words, you could end up marrying that woman, have children and live together, the rest of your life. This comes from the right words, at the right moment, to the right person .

The same is true for the Sacred name of G-d,saying it in a certain way, and incantation, it reaches the Creator's powers. If you do not know how to use it, at that moment, you direct unimaginable powers, that could cause devastation even destroying our whole planet. Remember our planet is a tiny grain of sand in the universe. The Creator's essence envelopes the universe, all the galaxies. They cannot exist without His essence. So, trying to call Him without knowing when or how, could cause irreparable harm. I will not expand on this subject. There is reference (Exodus 20: 1)

"All the people saw the sounds, the flames, the people trembled when they saw it, keeping their distance they said to Moses, you speak to us and we will listen. But let not G-d speak with us anymore, for we will die if he does." This was said to Moses, just after G-d said, the first two Commandments.

The other eight, G-d told Moses, and he in turn, told the Israelites.

So this gives you a small idea of the grandiose powers of the Creator, the Earth trembled and people fell'. The Midrash-מדרש said that they were dying and begged Moses for him to repeat the Commandments and not G-d. Moses wanted the Israelites, to hear first hand from the Lord. That's another reason that He spoke "You shall not believe the false prophets that will change my Commandments .

"Now, there are 72 sacred names of the Creator. They appear in some of the passages, in a hidden form. I expanded on this subject in my book, "Sapphire and Diamond-ספיר ויהלום', with G-d's will, I will translate part of it, because parts of the book will have no value in another language other than Hebrew, LASHON HAKODESH-לשון הקודש (the sacred language) the language of the Bible

The other Commandments are:

4) "Remember the Sabbath to keep holy."

5) "Honor thy father and thy mother."

6) "Do not commit murder."

7) "Do not commit adultery.

8) "Do not steal."

9) "Do not testify as a false witness against your neighbor."

10) "Do not be envious of your neighbor's house."

The fourth Commandment: I have already covered in past chapters.

The Fith commandment : We know the role of the parents in the world of Assiah-עשיה (deed), and above them is G-d, the All Mighty the Creator. In the physical world, they are the most important to you, just one grade below G-d. Do not forget, they were two of the three Energy-Intelligences that gave you life. They gave you, the physical body. The Creator gave you the life sparkle or (Soul), surrounding light, or Energy-Intelligence. Without it, the body has no life.

So your parents are the most important in the physical world. They are called in the Kabbalah, Lords to you. So if you disrespect them, G-d forbid, the Torah condemns, to spiritual death, the people who disrespect their parents. If you do, remember that you will be a parent one day. Your children will behave towards you in the same way you did towards your parents. They do what they see their parents do. This is called, "MIDA KENEGED MIDA" (measure against measure). Don't judge your parents, because you will be judged too, in this physical world.

The Slaughtering of a Kosher Animal:

What's the meaning of a Kosher animal? There is, in the Bible, a list of animals that are permitted for consumption. The most common ones are cows, sheep, and chickens. Before slaughtering a cow, the SHOCHET (or ritual slaughterer) checks the limbs of the animal, to make sure it has no broken bones, sores, wounds, infection or growth in the glands of the cow's neck. If everything is all right, then he prepares his long, sharp, unblemished knife. He then passes the blade over his finger nail to feel for any dent, if there are any, the knife is not suitable. So when he slaughters the cow by passing the knife and cutting the artery in the animals neck., he says a blessing. The cow won't feel a thing because the blade was so sharp, it just loses strength, and faints, but it feels no pain.

This is very important. Next, the cow has to be drained of the blood completely. When cutting open the animal, he takes the lungs, inflates them, and plunges them in a bucket of water. If bubbles come out, the animal had some kind of illness, the same is true for the heart, liver, etc., it is still Kosher, but not to the orthodox. If the lungs and all organs are good, then it is called Glatt-Kosher.

The rear part of the cow, is not used, but it is sold to the Gentiles because of the many veins and arteries and the blood in their interior does not get drained from the salt. After all of this process, the meat is put on salt for half-an-hour,then one hour in water. Only then, is the meat ready to be prepared.

You may ask why do all of this? First the ritual; the slaughterer must do it as a ritual, not take any pleasure in the killing or torture of the animal. The most important is, so the animal will not suffer. If it suffers, then it renders its energy perturbed, negative.The metabolism of the animal turns and causes the meat to be unfit for human consumption. So when we ingest the meat these energies, will be mixed with our own energy, and decrease it in purity regadless to the so many germs or ilnesses that could be in the blood.

If you take pure alcohol and dilute it with water, the percentage of alcohol decreases. In other words, quantity affects quality, especially if adding impure things. The same is true of the blood in the cow, all of the bad things are in the blood.

Example: people who have not committed a homosexual act, have contracted AIDS. If there is any contact with the blood of an infected AIDS patient, by an unfortunate child, due to blood transfusions, that child will get aids. Blood is the carrier of all diseases, and is the worst, for the human being.

G-d did not give us all those Commandments to harass us, to make our life impossible, but only for us to have a safe trek on Earth, and to increase our purity Do you think that G-D needs sacrifices of animals ? No, we do! the reason that young sheep or other animals that are under one year old and that have not been mated, were allowed for sacrifice. As I mentioned before, even animals, plants and minerals have Energy Intelligence and might have been a human in a prior life and are waiting for a chance for <<Tikun-תיקון >> (correction of the soul or karma).When we do ingest them, their energies are mingled with ours, thus given them an opportunity to reincarnate unto a higher form of Energy Intelligence i.e. a human.

That is the reason that we bless the food before and after we eat, thus during the ritual of the joined reincarnations into our body. Like parents who tell their children, don't do this or don't do that, they know better than they do, from personal experience, not to punish his child, or to bother him. The same is true of the Creator, He gives us the temptation and also the tools to fight them. Do not drink milk and eat meat together. The meat digests in 6 hours and the milk in a half and hour. This produces acids of spoiled lactate in your stomach, causing gases, bad odors, and your system to malfunction, that could lead to an ulcer.

Do not eat more than the body needs, negative Commandment No.195 "eating and drinking to excess", keep in mind that food is energy. What happens to a car when you overfill it with oil? 'The engine does not function adequately; black smoke comes out of the exhaust. The same applies to a human being releasing gas if he takes in too much energy (food), more than he needs, the engine (body) does not work properly and we would burn our energies inefficiently by straining the body and creating harmful gases.

Don't forget that machines and computers are made to function similar to a man. As exemplified by a car transporting us, our legs do the same, albeit much less efficiently. Computers are used to store memory like our brains however by different means. All of these are made in our likeness (intelligence). We create machines, and the Creator creates us to his likeness, intelligence-energy and Soul.

If we mistreat our body, he will render our life miserable, not performing properly. Like a car, if you fill it up with diesel, instead of unleaded gasoline, it will not run. The same is true of our food. Eat Kosher and you reduce the risks considerably. We even submerge all green vegetables in salt-water for half-an-hour in order to kill all of the germs on them, there are really many on them, and eating them, could cause you to eat an insect, it could be harmful to you. Most importantly, remember this, "We eat to live and, not live to eat.

"Another very important Commandment." You shall not tattoo yourself', because the surrounding light does not recognize those symbols, and could cause a confusion to the intelligence-energy, and your thoughts would be strange and to your disadvantage. Jewish people wear Phylacteries (tefilim-תפילים) that have 4 different portions of the Bible written on parchment which are in boxes made of leather. These boxes are painted in black on the outside, except for the part which we apply against the skin that is left natural. One is worn on the forehead, and another on the left arm. The strips of leather that hold the boxes in place are also painted black on one side, and left natural on the other that contacts the skin. The OR-MAKIF –אור מקיף or surrounding Light (Energy-Intelligence) can read the messages written in the tattoo and reinforce any wickedness, or deficiency inflicted on the perimeter of the body. So the tattoo confuses the Energy-Intelligence. On the other hand, Phylacteries have a positive effect on the Energy-Intelligence, and can restructure any deficiencies.

Our body is an illusion, our life is an illusion, if we could believe 100%, that everything is an illusion, we would understand our surroundings, and recognize that there is nothing really solid. Steel, walls, fire, are just a composition of atomic particles that are not glued together. Like a piece of wood or steel, it looks like one piece, but when seen in a microscope, the atoms or cells are close to each other, but without touching. They are a certain distance apart and are repulsing away from each other.

Stars function the same way, as if maintained by glue. The only thing they have in common is a surrounding light, that tells them that they are together, but they are also under an illusion. We all live in a world of illusion, and we work in unity.

Let's assume that you want to go through a wall. Without having an instance of hesitation and believing completely in it one hundred percent, you can go through. A person hypnotized, will go through fire without getting burned. It is because he is not in control of his mind, he won't get burned. The same is true for going through a wall. Under hypnosis, he will go beyond the reaches of the subconscious mind, and reach the inner energy-intelligence. It is Important that the hypnotizer, be completely confident, and not transmit to the hypnotized, any insecurity or uncertainty at the moment of contact with the wall, steel, or whatever.

If we develop control over our energies, we will finally comprehend the illusion of this world. As I have mentioned in other chapters, there are four very important elements to eliminate - anger, sadness, pride, and selfishness - in order to reach a higher level of awareness

THE POWER OF LETTERS AND NUMBERS

כחם של אותיות ומספרים

CHAPTER XI

The Power of Letters and Numbers

כחם של מספרים ואותיות

A lot of people ask the question, "Why repeat the same prayers every single day, Three times daily?" So I ask a similar question, "Why do we eat

three times a day'.?" The answer, if not obvious, is that we eat to stay alive, but only in a physical sense.

In view of this fact, that the body has physical needs, we often forget that spiritual needs exist as well, and nourishment is also needed.

Prayer is the food for the soul, that is needed to maintain a basic standard,and serves to balance the physical and metaphysical world of the body.

This is where the power of letters comes in. The book "SEFER YETZIRAH- ספר יצירה" is accredited to Abraham our forefather, it means, "The Book of Formation". He was the first astrologer. This is his explanation:

The Holy One, Blessed Be He, created the world with three things; writing, numbers and speech. There are 10 conservative numbers. Twenty-two letters in the Hebrew alphabet, three letters are the principals, seven double and twelve single. It is said, "G- d if you investigated me, you would know all." Because we humans are connected to G-d and we are made in essence. 'There is nothing we can hide from Him, as I explained in past chapters

The three principal letters are (Aleph א), (Mem-מ), and (Shin-ש) and the base has different elements in it that balance the language in the same way as a scale.

The seven double are (Beth-ב), (Gimel-ג), (Dalet-ד), (Chaf-כ), (Pe-פ), (Resh-ר), and (Tav-ת); the twelve simple ones are (Hey-ה), (Vav-ו), (Zayin-ז), (Chet-ח), (Teth-ט), (Yud-י), (Lamed-ל), (Noun-נ), (Samech-ס), (Ayin-ע), (Tsadik-צ), (Kouf-ק).

Against the seven double there are the 7 powers given to man; life, peace, intelligence, wealth, sperm (seed), grace (Charm) and kingdom. The twelve are the 12 elements that man needs: sight, hearing, smell, speech, swallowing, act of sex, walking, anger, laughter, meditation and sleep. And the meaning is that he separates the 22 letters in three part

Three letters:

(Aleph-א) the first of the letters beginning or initiation of words

(Shin-ש) in the middle of the mouth (Mem-מ) on the edge of the lips. The above letters are in this order in the alphabet, Aleph-א first. Mem-מ in the middle, and Shin-ש at the end.

Every two, are together and separate when a third comes to join one or the other, like in court, a witness comes to defend one or the other. The 7 letters that have a double tone. One hard and one soft.

Chanoun-חנון" = Charm, "Chen-חן" "Koah-כח" strength for government - (because government without strength cannot be a government). "Sheva-שבה" growing old, peace. "Zikna-זיקנה" old age = life - when a person arrives at old age, the storming of growing, lays to rest, chooses peace with his surroundings, and with himself "Praiseworthy is the man who fears Hashem(The Creator), who intensely desires his commands, mighty in the land will his offspring be, a generation of the upright who shall be blessed Wealth and riches will be in his house and his righteousness will endure forever. Even in darkness there shines a light for the upright," etc. (Psalms 1 12). "Here is the virtue of strength'valiant and sons', 'His offspring and life', 'generation and wealth and riches". (textual/literal). His righteousness, with it, peace as He said, "the act of charity is peace. (Issaih 32:17) 'the intelligence from the light', and piety is from charm.

"The Third Part: The twelve letters attributed to the twelve special strengths in man, within them, are the five senses and other strengths that are there for his convenience.

THE FOURTH THEORY:

3 Mothers with what is born from them. 7 leaders and their armies. 12 sides of the angles(the cube).

The sight is due to,Trustworthy witnesses". The world, the year, the soul and to each one the ten is drew. The 3, the 7, and the 12, they are in charge of the circle, the hunger and the heart. The 22 letters that 3 of them are the principals, 7 double and 12 single.

The three main ones (Aleph-א), (Mem-מ), and (Shin-ש) and to them a great secret, hidden incredible and fabulous. From them comes fire, air, water, and with them he created everything (earth was the center to everything).

The 7 doubles (Beth-ב), (Gimel-ג), (Dalet-ד), (Chaf-כ), (Pe-פ), (Resh-ר), and (Tav-ת) are said to be two types, hard and soft. We compare them to the strong and feeble and against them 7 different situations that change, Life and Death-Peaceful and Bad-Intelligence and Stupidity-Government and Slavery - Charm and Ugliness - Offspring and Desolate - Wealth and Poverty.

Also the seven: are seven and not six nor eight, and against them, the six, north and south, up and down, east and west and the center is THE CREATOR Himself blessed be He. He is the place of the world and not the world being His place. In Hebrew, the Creator is also called Makom (place).

The world needs Him for its survival, but He does not need this world in any sense or meaning. Also the winds. Four winds = up and down and He will always be in the center. Or the 6000 years of the world cycle, and on the seventh thousand year, rest.

Now the twelve letters. There are twelve sides which are equal in a cube. Where East is joined by the North. Another theory is that the world is counted asTen. Three are fire, water, and air, the Seven others are the planets (Mercury, Venus, Mars, Jupiter, Saturn, the Sun, and the Moon) and the twelve are the signs of the Zodiac.The year is also counted by ten; three are cold-heat, in between, not cold or heat. Seven are the seven days of the week and the twelve are the twelve months of the year.

The body is counted also with ten; the three are the head, the stomach and the rest of the body. The seven are the seven openings of the head; two eyes, two nostrils, two ears a one mouth. The twelve are the twelve functions that direct the body. Now the ten is also a triangle. (See figure No. 14 and 15).

FIGURE No.14: The PERFECT FIGURE

The two triangles that dividesthe circle onto 6 equal arches,similar to our world life cycle,onto 6 days of creation of 1000years each,the CENTER being the SABBATH OF CREATION. The same with the days of the week. Above all,this is a shield for the person or entity that will know how to use it. The main explanation is,that the two triangles show that on the top the CREATOR is ONE and only,and HE is also down in our world within us. And the other points are that He is around Creation,during all life cycle.

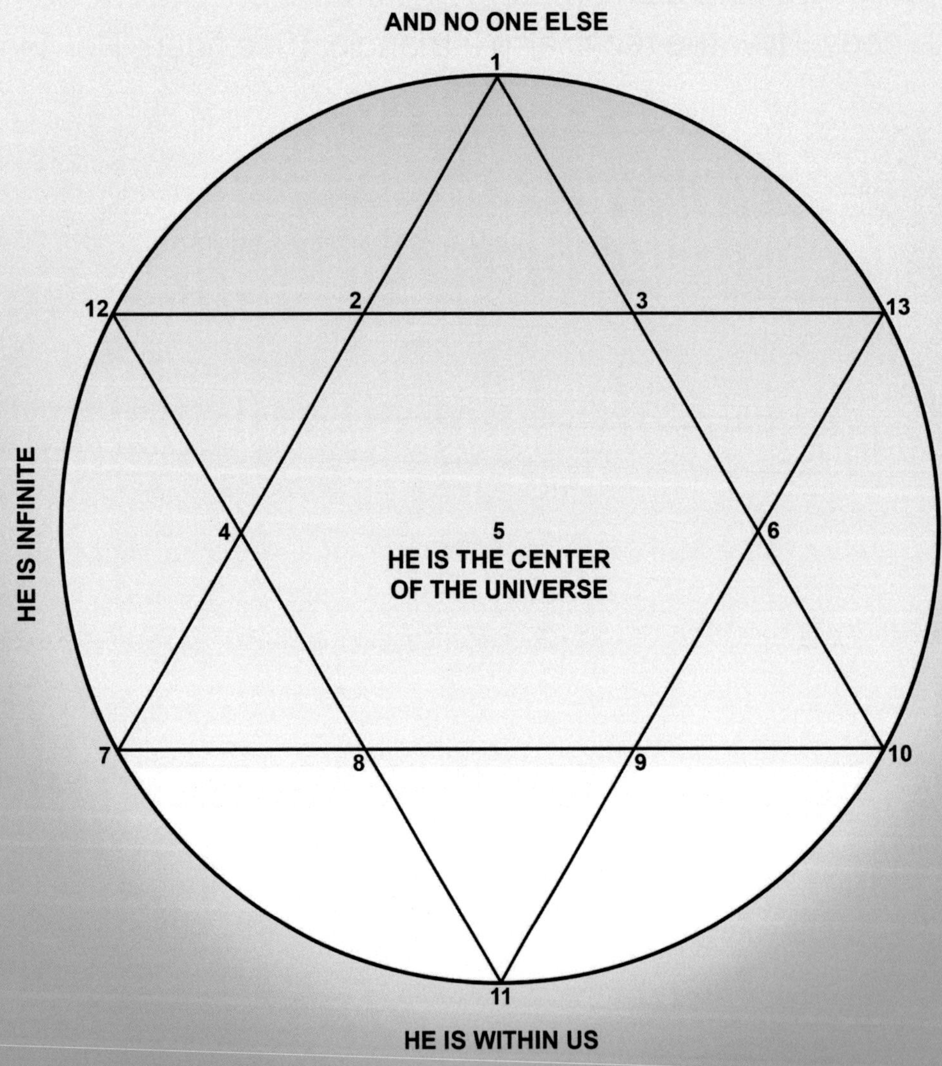

FIGURE n.15: The Triangle of our Spiritual World&Its Hierarchique levels.

This triangle shows Heaven's Hierarchy. On the bottom are the 4 components that are required to any soul to incarnate in our world.

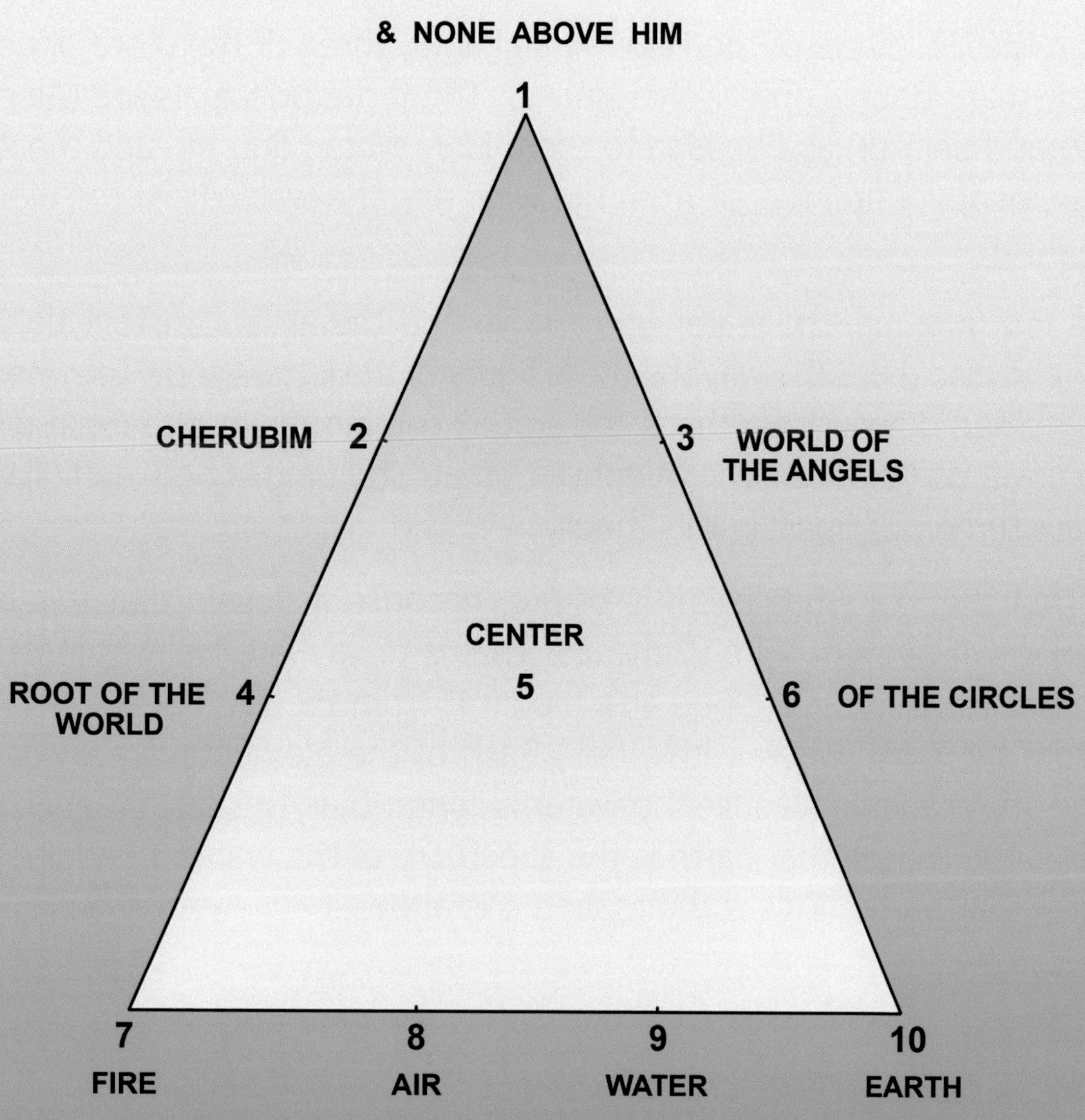

He is first and none above Him, or before Him and first in everything. The same explanation is the inner triangle that both form the Star of David, showing downward that He is also one in our world, within us.

The Fourth Theory:

From the 22 letters of the Hebrew alphabet, three of them are principals and seven are double and twelve singles. The three principals are (Aleph-א, Mem-מ and Shin-ש) as opposed to them Fire, Water, and Air, and the sky is from elements of fire, the air from the wind and the water from the element of the Earth.

The fire goes up, the water goes down and the air is constant between them - and also that's the way the three figure of the letters show the (Mem-מ) that is turned downwards (that's the water) coming down, the (Shin-ש) are facing upwards (like fire). The (Aleph-א), life, is standing in between (like air). Here are three principals in this theory;

1) That the sky (Shamayim) is of fire, because it is said that this is the chronology of heaven, "Fire"(Genesis). That is to say Fire/Heaven in reverse and in the store water dripped and the air is from the wind also from rivers and from the earth water within her.) Each of the elements is heavier than the one above them, because the water is lighter than the earth that is under it. This is the reason that earth is the sanctuary of the water. In addition, air is lighter than water. This is the reason it is above it

2) Fire is lighter than air and that is the reason fire is above it, becoming the highest of elements). The form of the (Shin-ש) is like the form of fire. It goes up.

3) The(Mem-מ) like the water coming down.

4) The (Aleph-א) standing between. Also the coupling of the letters that make words, or form them, is hard and soft. We say Emesh-אמשAsham- 1)אשם). The first is harder than the other. With Massa-משא Maash- 2)מאש), the second is sharper than the first. We say Shoam-שאם Shema-3) שמא). The first hard and the second soft.

5) These are the six rings, three hard and three soft. The same for the blending of the human and the rest of the creatures which forces the mixing of male and female species. If, when he finishes making love, the male's right testicle grows, he will have a male child. If the left one grows, he will have a female child. The same thing applies for a female virgin in her breasts when she becomes an adult. Additionally.

6) if the man finishes the love making first, he will have a female child. Conversely, if the woman finishes first, the child will be a boy.

7) This is only one of the six possibilities stemming from the fusion between a male and a female. Due to the blending of the two, due to the food that the couple eat before union, due to the sharpness of the father and the mother, due to intercourse, or due to the side (of the testicle) and also to the beginning and ending of the act (see above)

As I have explained in the third chapter, with regard to letters, however in this case they are not doubled in order not to separate the (Zayin-ז, Vav-ו,

Tet-ט, Chet-ח) - (Samech-ס, Kouf-ק, Tsady-צ), (Ayin-ע, Noun-נ, Lamed-ל) by themselves due to the multiplication that the main three letters and the seven letters to make the addition to the 22 letters of the alphabet. The fourth theory: 12 single letters. Here the Creator added, multiplied and totalled them.

The matter is as follows; from two letters of twelve, we can make two words. From three letters we can make six words. From four, twenty-four and from five letters, one hundred and twenty words, and so on.

Rambam-הרמב"ם Maimonides) said that if one by mistake wrote one letter on the Sabbath he is not guilty of transgression. One needs two letters to write or form a word. It is said not to write any "words" on the Sabbath. Not a "letter" - but it does not mean that one can write a letter on the Sabbath only to understand the concept, from two letters comes two words, as follows: x3=6, 6x4=24, 24x5=l2O, 120x6=720, 720x7=5040, 504x8=40,320 40,320x9=362,880, 362,880=3,628,800, 3,628,800x11=39,916,800.

We stop because the longest word we find in the scripture is of 11 letters (VEHAACHASDARPANIM-והאחרדשפנים) (Esther 9:3-אסתר) meaning (Satrap, Persian governor). Like the multiplying of the letters also will be for the humans and the rest of the creatures. The same way will be for the learning of science. The intelligence will multiply the same way and that's the way the brain functions in cross references; with billions of forming letters and numbers.

Like the parable of the king who wanted to reward his subject who saved his daughter's life. He told him to ask for anything he desired and it would be granted. So he asked him for a grain of wheat the first day. The second day for two grains, and the third day, for four grains and so on. Double every day from the past day for thirty-one days. If you ever try to calculate it, it comes out to over twenty million grains.

The Fifth Theory: The meaning is the ten names that we said that Moses engraved in his staff, Yah-יה, Hashem-השם, Tzevaot-צבאות, Elohe-אלהי, Israel-ישראל, Elohim-אלהים, Haim-חייםEl-, Shadai-אל שדי, RamVenisa-רם ונישא, Sochen-שחן, Vekadosh-Shemo-וקדש שמו.

The name YH-יה is made up of two letters and YH-יה VH-וה four letters.

The matter of TZEVAOT-צבאות, because it is a letter in His army. The matter ELOHE ISRAEL- אלהי ישראל that Israel are masters after him. Because he created nay thing that is called (CHAY-חי, " alive"), the liquid and the fertility. The liquid: The waters that were called (CHAYIM-חיים, life). The fertility and the tree that was called the tree of life MISHLE 3:1-משלי.

EL- לא He can do all, a strong power, strength and that is very complex, see (Barechi Nafchi-ברכי נפשי). M. Azulai, the system of poetry that the Rabbi Saadia Gaon wrote in Jerusalem. SHADAY-שדי, that His kingdom is infinite without interpretation too complicated to expand here, (see Sapphire and Diamond-ספיר ויהלם).

RAM-רם that He is Holy and His angels are Holy and praise Him every day with three Holy Holies, and it goes on and on. It is the most Holy secret studies of Jewish mystics. We could expand thousands of pages. When we say that G-d in heaven, because He is above our understanding, like the heavens are above our heads. He is everywhere all the time.

The Sixth Theory: It is said, the views of all the trustworthy witnesses, the world, the year, and the soul. On all of them, the base of twelve and seven on top of them, and three above all.

From the three elements He built His sanctuary.All of them are depending on one. Meaning to one that has no record. The only king in this world and that He is one and His name is one. All of this means that every twelve in the world, in the year and in the soul, are under the seven and the seven under the three. In the world the zodiacal signs (Mazalot) are twelve.

The influence of the seven planets are higher, and higher influence that the twelve zodiac signs. In the year, there are twelve months but are counted with only seven days, meaning that we count to a month.

The weeks are divided into weeks of seven days. In the man, the twelve organs that are in the head (two eyes, two ears, two nasal openings plus one mouth).

The base of the elements (Air-water-fire), that are really four. We already explains and it is from these three, that He created His sanctuary. Like it is said (in Psalms 26:8) Hashem, I love the shelter of your house, and the place of your glories residence."

The world in average, is His sanctuary. The small world, is the man. Because man is also a sanctuary to a spark of G-d. The soul is the reason to respect the body and life all together.

We say that the stars are in the sky. So are the eyes for the man, and the light in the candelabra in the sanctuary. It is said that the firmament is divided into two waters in the world. The curtain of the Ark (Parochet-פרוכת) separates from the Holy of Holies, and the Holy of the sanctuary. The partition is called the diaphragm. It divides between the organs of eating and breathing in man. The next of the 18 plains are parallels to those 3 that are really 4. The great sage the (RABA) explained that 18 things are in the sanctuary that is the middle world. The same in the higher world. It is also in the small world (the body of man). All of them are mentioned in Shemot (26:30 Exodus) and from that comes the word," Live" (CHAY-חי), which equals 18. That is the reason He said that from the 3 He built His residence. It is also said that all of them depend on the 1. That 1 is higher in number, because their joining are equal. If you add 3+1 it equals 4 and they are the square. If you add 5, it will be 9 and they are equal triangles. If you add to them 7, it will total 16, and they are square. If you add 9 to that, it will be 25 and they are also equal 5 by 5, and so on but the plains will not be added the same way.

The more you add, it will never come to the same figure. Meaning add 2+4=6, add 6=12, add 8=20, add 10=30, and everything you add on top of it will not come from them, what the multiplying factor was before, at all. Those form a different degree of separation, above them on the degree of unity, because we say one hundred, one thousand and so on until you say ONE world and above all the degree of unity because above it there is another degree of unity.

It is above all thought before it comes to be. Because, when it is discovered by the senses, it is without a doubt joined/parts called units. The real unity is when it will come up to the mind.

To every beginning without the senses discovering it. With that principal, we come to closeness of understanding. The Creator, will elevate in His Highness to One. He never came out of the matter of unity because He comes in our mind strong, steady and not a possibility. This is the reason we say: "G-d our Lord is one." But it is said, and His name is one. This will be in the time of redemption that all languages will be one.

When we pronounce letters or words, they form different geometrical forms in space (squares, circles, straight lines, wavy lines, triangles, octagons and so on ...). Every letter has its figure. When it is pronounced, it marks a combination of geometrical figures in space, that will stay forever in the Cosmos. Meaning that words do not disappear, although we do not see them. As it is said, "With every action there is a reaction."

This is the reason we should speak with wisdom, refrain from idle talk. All the words are witnesses to our behavior. They will be reviewed at Judgment Day, meaning the day we appear naked of body before our Creator. You will go to a higher realm, or come back again into the suffering of this world, or re-incarnate as a lower form of intelligence. For example, a person in a moment of anger, said "I'll kill him one day." It was just a way of expressing his anger. He really didn't mean it. Due to a reason stronger than his understanding, that person got killed. So, people who heard his comment, will remember.

Those words will haunt him, and perhaps, cause you to be accused of murder. Two words only could destroy your life.

The same in the metaphysical world. Weigh your words before you speak,prayers are important. They create shields and banners around us that protect us. They are energies in form of geometrical figures that stay in space and time. Do not forget, that we are a spark of the Creator, and we are His sanctuary. 'The soul that is a spark of G-d lays within your body and witness our behaviour; make your request in prayers. It depends on the Intention and Concentration of Faith, that our prayers are answered. They are the tools of the soul.

So really it is important to understand the principals of those theories, and comprehend the power of the alphabet and the numbers. could explain more and more, but the purpose of this book is to give you and insight of our universe, that could mean for us the 12 comparisons that I gave before, and we are the balance between Peace and War, Good and Bad, Beauty and Ugliness, and so on.

There is a Spanish saying that says, "You sleep in the Bed you have prepared." here is another reason that we repeat the same prayers every day as we do when you go to court and argue your case. You cite the same cases in the law that have won a similar case to yours. So, when we read the prayers we mention our forefathers and other tractates of the Bible that had the same plea and resulted in a good ending. So, you too can cite this <<cases >> for your own pleading (prayers) for G-D to grant you forgiveness or acceptance.

THE LANGUAGE OF DREAMS

פתרון חלומות

CHAPTER XII

THE LANGUAGE OF DREAMS

פיתרון חלומות

Reality or Illusion? Most of the dreams in one night are only reactions to the day's activities. But toward the early hours of the morning, 4-5-6 AM., our Intelligence- Energy leaves our body for a very short interval and connects to the other world. The world of dreams. This is very real and not an illusion.

The world we all live in is really only an illusion. Remember, just yesterday you were an adolescent. It seems so close but it is so far. The world, that we call the dream world, is very real and has its own language of symbols and images. I will proceed to explain .

The soul, or Energy Intelligence, leaves for that world. We are elevated to a different dimension,(depending on each ones capacity), to a higher level that we are able to reach. (See Figure No. 1 and 2). We see people who died long ago, who speak to us clearly. We see places that we have never been to. You see your future. Without knowing the language you will never know the warning signs. The other soul, or Energy-Intelligence on a different level is trying to warn or notify news to you. It is trying to tell you what is going to happen.if you could only know, you could avoid it. In some instances, the message could mean the contrary, in terrestrial language or in the world of Assiah (Deed).

Those departed souls, are in spiritual form. They are trying to communicate with you and show themselves to you in the way you knew them, in a language quasi-similar to yours. But beware because it sometimes means the opposite. Some people who die in their sleep, reach certain levels, that is hard for them to return in time. Because in earth time, it is seconds. In the other levels of awareness there is no time factor. Time is only a human definition and we live by it. We run against the clock. We look at our watches for everything.

I'll tell you a brief parable that would be a good example for what I have mentioned above:

There were two people who had no connection to each other. One was a normal person like you whom we will call Abe. The other, a mental-patient intern at a mental institution hom we will call Ben.

Abe was a hard working man. He worked for twelve hours a day and returned home very tired. His children, playing a video game, greeted him in an off-hand manner, immersed in their game, while Abe went to the kitchen to eat. He sees a note from his wife on the refrigerator which says, "there is a frozen dinner in the refrigerator, place it in the microwave, and I'll be home at 10 PM, Love, Daisy." That was Abe's daily routine, after eating, watching some TV and going to sleep before his wife returned. In the morning he wakes up, his wife is still asleep. He kisses her and leaves for work for another hard day.

On the other side of the city is Ben, who is very happy and thinks of himself as a prince, "the hospital" is his castle. The doctors and nurses are his servants. When he awakens, everybody greets him, and brings his medication, which to him are vitamins. He then goes for a stroll to the garden. Everybody salutes him. He is a lucky and happy man.

Returning to Abe, he is planning a trip to France with his family to Saint-Tropez. But all is riding on the deal he is planning, that if successful he will immediately inform his family, but he is so enthusiastic about it, that he tells them anyway before he has completed the deal. So everyone is preparing for the trip. The deal, as it turns out falls through, and he has no heart to tell them, and decides to borrow money and take the trip of his dreams even if it costs him dearly.

On the other end, Ben in his imaginary world, is also planning to go to Saint Tropez to have the same vacation so he tells his valets and majordomos to prepare his suits and swim- ware and to fuel up his jet which he plans to fly himself. The plane costs him 3 million dollars but he is very modest. He tells everyone that he was an aviator in W.W.II. Of course, all this is in his imagination.

Abe arrives with his wife and two children to Saint Tropez. The cab driver never returns his change for the ride to the hotel, so he has lost $15 that the cab driver withheld. When he gets to the reservations desk, there has been a problem and no room is reserved for four people, only for two. So they take a room facing the road and not the beach. He is beginning to get angry.

Then it turns out there is the Mistral (Storm) and he has to remain in his room. By now his temper is rising, as well as his blood pressure, to a dangerous level. He tells his wife that he is going out to buy cigarettes. The moment he leaves the room, he slips and falls down a twenty-two step stairwell and breaks both legs and one arm. He returns to the States, all in casts which cause him so much anger, that his blood pressure rises once again and causes him a stroke.

He is interned in a hospital and dies there a year later.

On the other hand, Ben in his mind, arrives to Saint Tropez, France. He is met by a limousine, at the airport by the owner of the hotel, with four beautiful girls who have heard of him and his exploits. The moment he arrives at the hotel, he is given the Presidential Suite with a beautiful view of the ocean, courtesy of the house. A big reception was there to greet him with the press present. He had a beautiful vacation. He returned to his palace. Everyone was there to greet him, and welcome him.

Of course all of this was entirely in his imagination only. The doctors however were quite worried that he was losing total reality . A year later he dies and is buried next to Abe. So, now to analyze who was the happy man, Abe or Ben? Abe was real, Ben, lost in his dreams. Who had a happier life? I leave you to decide. What is Reality?

Our life is compared to the shadow of a passing bird. What is reality? To you it might be a dream, to others a nightmare. In other words, we are in different levels of awareness. Each one of us is in a world of his own. One's mind is one's world. But we share the physical world. We have to abide by its rules, because we live in unity. Some are engineers, some clean the streets, some are doctors. We abide by a code of behavior, agreed by the majority. Therefore anyone who goes against the majority will find himself/herself in prison, or in a mental institution. Even if you are not dangerous or harmful to anyone, even not to yourself,but since you behave differently from the vast majority, you must be crazy. This is our world's rule.

From a higher level of awareness, our behavior will seem strange, because we assume certain roles. We are like the children who play doctor, but we play a role for grown-ups. This is also an illusion. Some play policeman, some judges, some criminals and so on. We are all playing a game of illusion. The day of truth will be the moment that will take off the suits of skin that we borrowed (Genesis) from this physical world.

As portrayed by an actor in different movies,he plays different roles. Once he is the good guy, and in another movie he might be the bad guy.

Now I'll proceed to explain the language of the world of dreams.if you perceive in the dream doing one of the below listed activities, it will mean you. If you observe someone else, it will mean that the person you are dreaming about is the one whom the dream translation applies to. Even though you are the dreamer, you are only being a messenger for that person.

Now if it is an Omen, there are certain prayers that should be said, or contact a Rabbi for "Hatavat Halom". Charity is also a good tool to reverse a decree. The explanation of the dream is composed of four parts: As colors, situation, person, and so on. To have a complete picture, you have to research all the components of your dreams.

Again, the meaningful dreams are those ones you dream just before you wake up in the early hours of the morning only. The ones you dreamed before that,they are only reflections of your days activities.

WITH THE ELEMENT OF EARTH

INCLUDING COUNTRIES, FRONTIERS AND PLACES

a)Who sees himself standing on a mountain or scaling a mountain;He will grow high,in social position.

If he comes down from that mountain; He will therefore reduce his social standing.

b) The one who percieves in a high place; High placed people in government,will place him in a high esteem.

Goes up on a ladder; Very good for him.

Looks down from a high place; Long life he has been prescribed

c)Goes up to the roof; Raise in position(business or social)

Comes down from the roof; Lowered from his position(business or social).

d)The one who excavates the ground; Not to worry(in any situation). Even if he is seen digging a grave.

But if He is seek; (in real life);It is an omen.

e)Seeing oneself carrying earth; It is a presage(a hint to a grave).

The one who perceives in his house or in his country: from a danger, he will be saved.

That he has been arrested by an officer: a protection is being decreed from any harm.

And if he is chained: he will have a higher protection (from a high level of awareness in every instance).

f) If he stands in a narrow place: Suddenly he will earn a good profit and it will be good for him.

The one who observes in a house embroidered or drawn; Happiness and peace is predicted.

The one who discerns,hanging a door at his doorstep; He/She will marry soon.

The beams of his house breaking; It is an Omen.

The frame of his house broken; His wife and children will die soon(G-d forbid).

If an old home is being destroyed; It is a good sign for him.

If a wall of his home falls; Death and trouble will happen very soon.

If the walls of his home fall; A son of his will die(G-d forbide).

g)He who sees himself in the east: An enjoyment will eventuate.

He who perceives himself in the west; He will travel to a far country and he will return.

If he descerns in Egypt; His good times will increase.

If he sees himself in Israel or Jerusalem; The creator will make him a great personality over a country.

If one perceives in Jordan; His share is with the righteous people.

If he sees himself in country after country; He will travel frequently from place to place.

If he returns to the country; His good times will increase.

h)The one who sees himself in a country that the name starts with the first letter;

Aleph א A = It is good for him

Beth ב B=He will build a house.

Guimel ג G=G-d will grant him A charm and grace.

Dalet ד D=A banner of grace is over his head.

Hey ה H=G-d prepared him.

Vav ו V=Woe to Him.

Zayin ז Z = G-d granted him goodness and wealth.

Chet ח CH = illness he will endure.

Tet ט T = It is good for him, he should go to another country and he will become rich.

Yud י Y = G-d will cover him under his wings. (Under His protection).

Chaf כ C = Above a preparation has been made for him.

Mem מ M = His livelihood is ready and his enemies will fall.

Noun נ N = His destiny will be completed with good luck and profits.

Samejh ס S = G-d is relying on him because He loves him.

Aayin ע AA = The eyes of the Creator are looking after him of the good for the rest of his days.

Pe פ P = G-d will save him,and he will receive only good for good for the rest of his days

Tsady צ TS = He is righteous and honest.

Kuf ק K = G-d is close to him, because He regards him as good

Resh ר R = He will attain all his hearts desires.

Shin ש SH = G-d will listen to all his prayers.

Tav ת T = Honest and sincere.

WITH THE ELEMENT OF WATER

a) Who sees a river flowing calmly~: He will have peace.

Who sees a river flowing turbentlv: He should fear his enemies and bewAre of them.

b) Who sees a well: He perceives peace, some say he found life.

b) Who sees rain: Gains he should anticipate.

Who sees torrential rain falling out season: A tragedy will happen to that country.

Who sees torrential rain falling in season: Greatness will come to that country.

Who is standing in the rain: Good will happen to him.

d) He sees the sea corning to him to drown him: He will fall into his enemies hands, and they will harm him.

He perceives walking toward the ocean: It is better not to take a walk that day, to no place at all.

Enters a pond: He becomes head of an institution or school.

Enters a forest: Becomes a teacher or tutor.

He drinks rain water: A joy he should anticipate.

He drinks hot water: It is good sign

He drinks cold water: His wealth will increas

He drinks the water and it is bitter: Presage of a fight.

Who sees a bathhouse: Will be saved from a trouble or problem

Who bathes in hot water: Misfortune will befall him

Who bathes in luke-warm water: A joy will happen

Who bathes in a river: An omen

Who bathes in a spring: Good things will happen

Who plunges into the ocean: Peace will come

Falls in a river: A good profit he should anticipate

Who bathes in cold water: His misfortune diminishes and good news he should expect

Enters a bathhouse: He will be saved from bad people

Who perceives as the owner of a bathhouse: He will attain high social position and astonish people in his surroundings

Bathes in a bathhouse: A misfortune will befall him

e) Who comes out of a bathhouse: He will be saved from a misfortune

Who swims in the ocean: Peace he will attain

Who swims in the ocean and the waters are turbulent: He will fight with his bosses

He who navigates in the ocean. pool or river: He will befriend good people

If he drank from a spring: A very good satisfaction he will have

Navigates in mud and drowns: A misfortune will befall him, and he will be saved.

Some say he navigates: He will be saved from any misfortune or trouble

He passes over mud & mire: A fight will take place

f) Who sees himself drowning in an ocean or befoul in mud & mire: Sickness or loss of money he will endure

Fell in the ocean: A sickness will befall him, and will be saved.

And if he drowned: He should plead for mercy because he might die

And some say if he fell in the ocean or a river: Presage of a mourning.

WITH ELEMENT OF WIND

The wind carrying him: He will attain rule, authority and power.

Metals

a) Who sees an entire coin: Wealth he should anticipate.

Who sees pure crystal or quartz: G-d loves him.

Who sees a pearl: Honors and government he will obtain.

A seal or ring without a jewel: He will do a deed for which he will be praised for.

A ring with written letters in silver: He will rise to a great position, until strangers will envy him.

A ring on one of his fingers. except the small finger: He will become a suspect.

He receives a ring as a present: A very good sign for the dreamer

b) Who sees a ring of silver or tin: Good sign for him.

Who sees a ring of steel: Better and suddenly he will gain big profits.

A crown of silver or gold if he is an artist: Good sign.

A crown of silver or gold if he is not an artist: An omen.

All kinds of gold: Bad for him and in the future he will have troubles with people.

All kinds of silver items: Good prediction.

c) Who sees money and gold coins: Profit will materialize.

He who finds a coin of a treasure: A great satisfaction he will have.

He lost them and found once again but broken: A misfortune will befall on him.

Found utensils of silver and gold: Honors he will attain.

Found brass or rings of brass or copper: He will have an argument

Found gold or pearls: He will become important to a president or king

Found a copper penny: He will have an argument.

Found a finding (if he is rich): He will become poor..

Found a finding (if he is poor): He will become rich

d) Who sees that utensils are of silver or gold: All things that all his enemies thought about him will be found to be groundless.

Found tools made from steel like nails or spikes: He will have a quarrel with a big Family.

All kinds of metal: Are good for the dreamer.

ELEMENT OF FIRE

a) The fire that is not made to serve but to burn or destroy: Which evokes evil Made under controlled substances like candles and the like: Subject matter of the Dreams.

b) Perceives a small oil lamp, lantern. & lighter in his room: He will remarry.

c) As above but extinguished: G-d forbid, an Omen.

d) Fell into fire: Will have the Creator's will.

e) His home burned down: He will have an argument.

f) Walls of glowing coal: His body will be missing.

g) Sees coal growing dim: A quarrel with haters.

WITH PLANTS

Who sees himself in a planted field: Good times he has been prescribeb

Who sees himself in a planted field of plants that are edible: Good for him. has a garden and eats vegetables from it: A bad reputation he has been credited. Some say his wife is cheating on him.

Who sees wheat germ: Has seen peace.

Who sees barley: Profits, silver and gold are anticipated .

Who sees in his hand a germ of green wheat: It will generate profits.

A cabbage or other cooked vegetables: Much good and riches he will have

Who eats a loaf of pure wheat: He is promised admittance into the world to come.

Who eats a morsel of semolina: His prayers will be surely hear.

Who eats bread with vegetables: A bad name was attributed to him .

Who sees that he holds flour in his hand: Profits he will obtain

Eating all kinds of vegetables at the table: Good profit and good life he will have.

Eating fresh onions or leek: He will endure pain.

Eating spinach: He will receive much good and riches from the Creator.

All kinds of are good to the dream but the heads of turnips called germana and certain say also onions and garlic.

Who sees horseradish or other bitter herbs: An Omen.

Who sees himself cutting ears of corn: Profit he will receive.

Eating vegetables in vinegar: It is a presage.

Who eats pumpkin and squash: All kinds of sickness will befall on him.

Who eats green peas: Profits he will receive .

TREES

a) Sees that he plants trees: A good deed, has been credited to him .

He planted a vineyard or sees trees flourishing: Good things will come to him.

Who sees a fig tree: His theories are being kept.

Who sees a myrtle: His assets are growing, (if he has no assets: He will receive an Inheritance).

Who sees a plant that grows fruit: It will be good for him and his wife.

Who sees a palm branch: He has only one heart, for his father in heaven.

Who sees a loaded vine: His wife will not have a miscarriage

Who sees a choice vine: He can expect the Messiah (redemption).

Who sees cinnamon: Wisdom, he should anticipate .

Vineyards and date palms: Denotes a woman

Date Palms: Denotes a respected woman.

Who sees himself walking into a vineyard that does not contain any grapes: Anticipates mourning.

Who sees a tree full of fruit: Profits are anticipated.

Who sees fruit on a tree: He will rely on an important man and he will enjoy his Presence.

Who sees a tree with sweet apples: It denotes a good and rich man.

Who sees a tree with sour apples: Denotes a bad and sour man.

Who sees a nut tree: It denotes a hard man, who is, bad company .

Sleeps under trees: He will have sons.

Sleeps under a tree: G-d will come to his help.

Climbs the tree: Demonstrates respect.

Walks around trees: He will have sons.

If they fall: His sons will die.

Uproots a tree that bears fruit: Bad things will happen to him

Cuts down a tree: Bad for him and his home.

If he sees trees being pulled: War and panic.

If he sees tall trees: Abundance and peace.

Who stands against a wood pole: Goodness to be expected

Who sees wood supports of beams or columns made from wood that will support a house with the top or bottom broken: An omen, he should quickly give to charity, he might be saved .

FRUITS

Who sees the first fruits of the year: Something great will happen to him.

Who sees the first dates of the year: Good, but not for the government or any official. Another explanation; his sins are canceled.

Eats dates: A misfortune .

Eats almonds: An illness he will endure.

Eats almonds that are stale or sour: (same as above).

Eats pears and or Quinces: Will live a good life.

Who sees white grapes in season or not: It is a good sign.

h) Who sees black grapes in season: It is a good sign.

i) Who sees grapes and eats them: He is promised to be in the world to come.

j) Who sees black grapes not in season: Bad Omen, needs to pray for forgiveness.

k) Who sees grapes that are red, and eats them: Good for him and his children.

l) Eats grapes: He should not go in front of a judge or authority, because his enemies will defeat him.

m) Sees raisins: He will receive profits.

n) Sees soft figs: He will make money in business.

o) Sees dry figs: He will have profit.

p) See black figs: He will have to do much work, labor .

q) Who sees small pomegranates: A lot of business like the grains in the pomegranate.

r) Pieces of a pomegranate were taken: If he is a scholar, he should expect the Torah; If he is a complete ignoramus: he should expect good deeds.

s) Who sees olives: Multitude of good business.

t) Who sees an olive tree: He will father boys.

u)Who sees an ETROG (citrus): He becomes respectable and honorable in the eyes of his clients .

v) Eats a lemon: He will become poor .

w)Eats nuts: He will have a good life.

x) Received nuts: He will have a good life.

y) Eats apples and oranges: A great life, and a person of great merit in it.

z) Sees red crab apple: illness will overcome him.

aa) Sees white crab apples: He will have wealth.

bb) Sees peanuts dry or soft: All his troubles have been canceled cc)

Sees peppers: It is a very good sign.

dd) Ginger or ZANGWILL: His name is in every country.

All kinds of fruits are good for his dream except unripe dates.

FRUIT JUICES

a) Who sees a wine press: A quarrel is anticipated.

Who sees a wine cellar: Rain is forecasted.

Leaks on top of a wine cellar: A quarrel he will have.

Sees grapes being pressed: Goodness he is assured.

Who sees himself pressing grapes and drinking from them: He will have wealth.

Who sees a barrel of wine: A great joy will come to him very soon.

Who sees himself as a guest of people at a bar: illness he will endure.

And if he doesn't drink with them: He will be saved of his illness.

Who sees himself in a bar: He will suffer losses.

Who drinks wine's ingredients: A great pain he will suffer .

Drunkenness: Shows greatness.

Who sees himself drunk: He will be in great trouble, but will be saved.

Who is a drunk: Profits are anticipated.

Sees himself drinking: He should be aware and keep himself out of anything because he might fall .

All kinds of drinks: Are good for the dream (but wine).

Who sees olive oil: He should expect illumination from the Torah

Drinks oil: His wife cheats on him, and so does he.

Who sees himself drinking alcohol made from grain: Health and good money he will have.

Who sees himself drinking alcohol made from a crop: Good livelihood and wealth he will get.

Who sees himself drinking alcohol made from dates: Good wealth he will have

SHIPS MADE FROM WOOD

a) Who sees ships: Abundance, money and power.

b)A ship and he is the duty officer on the deck or that he falls in the ship: He will be saved from many misfortunes or troubles.

c)Enters the ship: He will go to a far away country

and some say he and his children will be in the world to come

And some say if the ship is small: He alone will go to the world to come.

d) Who sees himself as duty officer and the ship navigates as if by a miracle: His fortune and position of greatness will increase.

e)Being in a ship and it is docked in port: He will anticipate for good, very, very much.

f)Sitting in a small improvised raft at sea level: A good name he will attain, also a great reputation he will gain for himself and his family.

g) Who sees that he fell into the ship:G-d will save him from his trouble.

h)Fell from the ship to the ocean or river: He will fall into his enemies' hands, and they will harm him.

i) Who sees himself sitting in a small ship: Good for him, and he becomes the head of his town

j) Navigates in a ship even though it is not his habit to do so: He will go to jail.

k) Who sees a ship that sank: He will be saved from his sins and from purgatory.

l) Who sees a ship that sank and he is saved: His sins are pardoned, and all bad thoughts that people had of him, will be annulled.

m)Who sees that he fell from a ship onto dry land: Good prediction

n).If he comes out of a ship onto dry land: He will discover many countries.

ANIMALS

FIRST PART: PURE (KOSHER)

Who sees a bull: He Will acquire a good name.

Who sees bulls or sheep: He will rise to greatness.

Who rides on a black bull: He will achieve fame.

'Who sees bulls or cows: He will see the downfall of his enemies'.

Calf eating in pasture in a field: Joy and profit will come his way.

And if they are asleep: Laziness and hard times are anticipated.

Who sees a bull pulling a cart or not going on the road: He will hear news that will make him happy.

Riding a bull and then falling from it: Death will occur soon.

Bull riding on him: That same year he will die.

If the bull kicked him: He will that year travel far and wide.

Drank from her milk: Great satisfaction and will have a good year.

If he bit the bull: He will lose something on the way.

If the bull bit him: He will live many years.

If the bull trampled him: He will have a serious problem.

He rides the bull and the bull is going down a road: Troubles he will have.

Eats from the bulls flesh: He will be rich.

And if he is rich already: He will become richer.

If the gull gores him: He will have sons.

Who sees a goat: A year has been blessed for him.

s)Who sees goats: Years have been blessed for him.

t)Who sees a flock of goats and he is the shepherd or is buying some of them: His sons and daughters will multiply.

u)Who sees many he goats: Rain will fall and it will be a blessing.

v).Going at him: He will raise to greatness and will be very helpful to people because of it.

w)Drank milk from them: His livelihood will increase.

NON-KOSHER ANIMAL

Who sees a camel: A death was decreed for him and he will be saved.

Who sees camels. It is an omen.

If he's bitten by the camel: His luck will change for better.

If he rides the camel: He will travel that year.

If he falls from the camel:He will suffer a great illness.

If the camel kicked him: He will hear great news.

If he killed the camel: He will kill his enemies.

Eats from his flesh or drinks from her milk: He will become very wealthy.

Who sees many horses and it is the month of(Tishre=SeptOct):In the same time he will die.

Who sees a white horse: It is a good sin .

Who sees red horse: It is an omen.

Who sees a red horse standing still: It is a good prediction

Pursues a horse, reaches him and strucks him: He will win over his enemies.

Pulls a horse behind him:He will receive advice from great people.

If with a bridle: A great man will be hearing his advice.

Riding him or stolen from him: He will lose his merchandise.

Mounted on him: Greatness and good, he will have.

He falls when the horse is running: Death it is.

He falls when the horse is going slow: He will build on his sins.

Eats from the horses flesh: G-d will resolve his financial situation and his money will be safely guarded .

All horses are good for the dream except red ones!

Mounts a mule: A misfortune will befall him.

Fell from it: In that same time he will die.

Eats from his flesh: He will take forbidden money.

Who sees a white donkey: Good will come soon from his wife or he will receive aninheritance.

Mounts a white donkey: He will anticipate redemption.

Works the ground with it: A loss at home.

Who sees the king's mule or donkey: Greatness he will have.

Who sees himself on a donkey without a pack -saddle: He will get ill or lose his money.

Who sees himself on a mule with a pack-saddle: A blessing, he received.

Who sees a donkey kicking him or soiling him with mud: He will receive a present from a great man.

He rides the donkey with a harness: He should expect the Messiah.

Falls from him or eats his flesh: Bad things will happen to him.

Sees a pig: He gets a good livelihood opportunity.

Eats from his flesh: Will add riches to his wealth.

Rides on him: Beware of his enemies.

DANGEROUS BEASTS

Sees an elephant without a harness: A marvel or wonder will happen to him.

Elephants: Wonder of wonders will happen to him.

Mounts on him: He will rise to greatness and power: .

Lions: Long life he will live.

A group of lions: He will have a good year.

A lion roaring at him: He will endure illness and misfortune that year.

Fought a lion and won: He will defeat his enemies.

A lion pursuing him: His enemies will ambush him, and are trying to get him.

Became friendly to one of them: He will make peace with his enemies

Sees lion cub before him: Many enemies will join forces against him

Bites one of the cubs: He will win over his enemies.

Eats from the cubs or lion's flesh: He will have a great fight.

Who sees a bear pursuing him: A man will overcome him.

Who sees a bear's head or eats from his flesh: He will receive forbidden money.

Sees a bear approaching: A quarrel.

Sees a wolf: An enemy will rise against him very soon.

Sees a snake: Will have an increased livelihood.

A snake bites him: His livelihood will be doubled.

Killed the snake: He lost his livelihood.

The serpent escapes from him: An enemy will rise against him very soon.

Sees a serpent asleep or curled around his neck: Livelihood expected.

Sees a serpent with his mate: Wealth will be attained.

Sees a serpent in his lap: Will have an increased livelihood.

A snake in the water: If he has no wife); He will marry.

A snake in the water: (If he has a wife); He will be widowed.

THE MEANING OF DANGEROUS BEASTS: ENEMIES

a) Sees a Ram or Deer pursuing him: A fright without bad effect.

b)Kills a Ram or Deer: He will spill innocent blood.

c) Rides one of them: A misdeed is credited to him .

d)Someone gave him one of them as a gift: Good and greatness he will have.

e) Sees dogs running: Slander is said about him.

f) Plays with dogs: He will fall in love with his enemies.

g) Barks with dogs: A thought from his enemies will rise.

h)Dogs pursuing him: He will run away from the authorities of that country.

i) Sees dogs barking behind him: His enemies will ambush him.

j) If the dog bites him: His enemies will win.

k) If he bites the dog: He will eat with his enemies.

l) If he is on top of a dog: He will overcome his enemies.

m) The dog is on top of him: His enemies will overcome him.

n) If he eats form the dog's flesh: He will be moved to a safe place, that is good and peaceful.

o) Sees a fox: He will be sick.

p) Others say a fox or a rat: He will have a good year.

q) Sees a cat: He will receive beautiful clothing.

r) Some say if he sees a cat: A certain death will be decreed and he will be saved.

s) He who sees a mouse and a rat: Indicates a new face.

t) Desert beasts: Unnecessary trouble looms.

u) Starved beasts: Trouble and illness is anticipated.

v) Trampled by angry beasts: Lies will be said about him .

w)Standing next to a beast: His next of kin is his enemy.

x) A dead beast: His enemies will die.

KOSHER FOWL

a) Sees a gander: He will expect wisdom.

b) He who sees a gander in his home: He will become "Rosh Yeshiva" or head of a School.

c) Fowl flying: Greatness he will anticipate .

d)Sees a rooster: He will father a son(s).

e) Chickens: A nice apartment he will get.

f) A lot of chickens that are going to his home or in front of him: Great wealth and respect he will have.

g) A castrated rooster: He will not have children.

h) A maid selling roosters: He will father baby boys.

i) Sees a dove or baby pigeons and he catches one of them: He will have baby boys.

j) Catch a pigeon: Will father a baby girl.

k) Tied one of them: He will receive an inheritance.

l) Eats from their flesh: His luck will be sustained.

m) Sees a bird: He has seen peace and it is good for him.

n) Sees a bird in his hand: He will receive good news.

o) Captures the bird that ran away: He should fast that day and give ch.

p) Eats one of them: It is a good prediction.

q) Sees roosters or all kinds of fowl fighting: He will quarrel.

- NON KOSHER FOWL

Sees an eagle flying: He will rise to greatness and/or will be wealthy.

Capture an eagle: He will become rich and defeat his enemies.

Killed an eagle: He will govern people.

Eats his flesh: He will be needed for mankind.

Sees raven flying toward him: It is a good prediction.

Sees raven on his head: He will die.

On the head of another person: That person will die.

Capture a hawk: Good and honor will he get.

a) He who sees a vulture or beast on top of his home: He will become ill and recover.

Captures one of them: He will have an argument with someone who trusts him .

Killed one of them: A good thing will happen to him.

Eats from their flesh: Many good things will happen to him .

Sees an owl or a pelican: He will get ill and recover .

Sees one on top of his home: He will mourn.

Killed one of them: A financial loss.

Heard his sound but didn't see him: It is a good prediction.

Eats from the bird's flesh: He will suffer a loss.

One of them bit him or he lost them: Good and greatness is expected.

Who is pursuing an ostrich and could not capture her: He will pursue wealth without attaining it.

Mounted on her: He will raise to greatness.

Sees bees: His enemies will rise against him

FISH

Fish: Will make money working.

Who takes fish from people: He will take money with disgrace and shame.

His fish are missing: His money will be missing.

Caught them in his net: He will father boys.

Fished them with his fishing rod: The Creator will grant him a good life.

f) Sees that he caught swarming or reptiles in the ocean: An omen Dolphins and all the rest are good for the dream, either on the ocean or the land.

Sees small fish: G-d will give him gains.

Sees big fish: G-d will give him big gains.

Caught them: He will receive good news from another place

PART A: MILK AND CHEESE

Sees breasts full and suckled from them: Gain is anticipate

Breastfed from women's breasts: Big gains.

Milk from sheep or goats: Wealth he will have and his luck will change for the good .

Drinks milk from deer or ewe: He will father a boy and should expect a lot of good.

Drank sheep's milk: Great gains.

Drank milk from a mare: He will find money.

Drank milk from animals: If sour, a presage; if sweet, good

Who sees that he eats fresh cheese: Gains it is.

Butter of goat's milk: Greatness he will have.

Other kinds of butter: Good news he will have.

Sells cheese or milk: Greatness he will have.

PART B: EGGS

Who sees eggs. nuts or crystal items and other things that could break: His request is pending.

If they broke: His request is being fulfilled .

Some say if he sees eggs: His request is fulfilled, and credit he has.

If they are cooked. boiled or broken: His request will be fulfilled

Eat them cooked: He will be saved from an upcoming illness.

If he eats them hard boiled or broken: His prayer has been accepted, some say hard boiled A good life he will have.

If he eats them fried: An omen.

If he eats them raw: His grace before G-d will be increased.

All eggs are good for the dream, except Ostrich egg

PART C-HONEY

Who sees honey together with bees: He is surrounded by his enemies.

Uprooted a root of a tree, and honey poured from it and drank from the water: He will be saved from misfortune .

Eats honey: Bad prediction.

Some say: Good honey: Good prediction.

Some say: Bad honey: An omen.

Eats cooked honey. and it is sweet and dry: He will become ill

Wafer or cakes made with honey: It will be good for him.

MEATS

Sees himself eating a roast: A misfortune will happen to him

Sees that he eats a cooked meal or raw. and it is sour: It will be bad for him.

Eats some that is salty or dry: A serious illness will occur to him.

If it is sweet and without sauce: A joy is announced.

Eats in a bowl and cleans its contents: His luck is departed from his world.

Drank blood: He will be saved from misfortune .

ON MAN AND THE HAPPENINGS TO HIS BODY

Head. intestines and penis or middle finger: Shows the dreamer.

Shoulders: Women or sisters.

Arms: Sons or daughters.

Forearm: Friends or relations living at home.

Hand: Servants.

Testicles: Sons.

Buttocks. feet and legs: To all life's existence in the dream

Thigh: Relatives.

Loins: Selfishness or riches.

Blood and livers: Indicates a treasure.

Some say liver: Desire.

Blood: Anger.

Heart: Life.

Infections with blood or pus: Will indicate gold.

Hair of the beard: Indicates beauty and frailty.

Hair from the forearm: Indicates women.

Pubic hair: Enemies.

Rest of hair: Riches and beautiful items of gold.

Flying: Indicates change of place, change of situation like seen from a flight view.

Sees himself tortured or fasting: It is bad for the dreamer, and it is recommended to fast even if it is the Sabbath.

Sees that he became taller: He will live a long life.

Running: Good for the dreamer.

Sleeping: Good for the dream.

Sees himself sick: He will rejoice that same year.

Sitting on a man's neck: He will dominate his enemies.

Fell from a handrail or judgment in court and won: A misfortune it is.

Washes his head: Saved from any misfortune.

His forehead is broken: Misfortune it is.

Something in his eye: An omen.

Dead: Good prediction.

Hanged: He will rise to greatness.

And if he loses his head: Do not be afraid, your hour is delayed, repent,

He cut his nose: Anger abandons him .

blood from his throat: He will be ill and will recover.

Strained with blood: Lost his money.

He lost his molar teeth: His daughter or sisters will die.

Who sees that his teeth grew: A great salvation will occur.

A growing tooth: An illness is predicted.

A blackened tooth: A misfortune is around the cor- ner ..

A moving tooth: An illness is anticipated.

A tooth that is loose and then falls out: Death is predicted and he should fast even on the Sabbath and pray for forgiveness And many say that

if it is hurting when it falls: A person will die and he will be devastated.

His jaws fell out: All bad advises against him will fall.

Sees himself mute: Goodness is predicted.

His beard trimmed with a blade: An Omen.

With Scissors: A good prediction.

His beard is uprooted: He has gained a bad reputation.

Completely shaved his head: Good is predicted .

Who shaved his head in a dream: It is a good sign for him.

Who shaved his beard and his head: It is a good sign for him and for his family.

He bleeds: His sins are forgiven .

Bleeds from his nose: His days to live are fewer.

Bleeds from his shoulders: He will lose big.

Bleeds from his arms: A small loss.

Sees himself purified: An omen.

Someone wounded him: His days will be shorter to live .

Bled from his body and it did not trickle: His sins are remembered.

Urinates blood: His wife will miscarry.

Sees his arms pure and white: He will fall in love with a great person.

Sees his arms ugly and dingy: The people he loves will speak falsely of him.

Swinging in his dreams: It is a good sign for him.

Urinates on a utensil made from silver or gold or

something of value: He will father with a honorable .

Sees himself barefoot: A loss is predicted.

Sees himself legless: He will travel to a far away place.

One of his thighs or knees cut-off: He will become very sick and recover.

WITH GREAT PEOPLE:

Sees a king: He will rise to greatness.

Speaks with him: (If he greets him) good; (If he speaks in anger to him) An Omen.

All ministers are good.

Sees a slave: Bad presage.

Sees a woman slave: Good is predicted

Sees King David: Righteousness is anticipated.

Sees King Solomon: He will expect wisdom.

He sees Rabbi Akiva: He will expect righteousness.

He sees Rabbi Azaya: Greatness and riches are anticipated.

He sees Rabbi Yishmael Ben Elisha: He should worry about troubles and calamities.

He sees Rabbi Ben Assai: He will expect righteousness.

He sees Rabbi Zoma: He will expect wisdom.

He sees someone else: He should worry about calamity and troubles.

Ismael son of Abraham: His prayers have been heard.

Pinchas: A wonder will happen to him.

Rabbi Huna: A miracle will happen to him.

Hanina, Hananya, Yohanan: Miracles will happen to him .

Sees a man whose name has (Shin) a SH: It is bad for him.

Sees a man whose name has (Nun) an N: Like he sees divinity himself.

If the first letter of the name is (Yud) Y: An omen.

CHAPTER ON THE DEAD

Sees himself sneaking to the angel of death: He will get ill and recover.

If the angel of death is standing at his head: He will die.

If he stands at his feet: He will sicken until almost dead and then he will recover.

Who perceives dead: He did a good deed that makes him closer to G-d, some say; years have been added to his life span.

Sees mortuary items: Same as above.

Who sees himself being buried: He will be handed to a cruel person.

Slept in a cemetery: He will spend a night in jail .

A dead at home: It is a good sign for the house.

If the dead eat and drink at home: It is a good sign for all the people living at that home.

The dead took a shoe or sandal and went out: It is an omen for the house.

Sees the dead: if he is healthy: He should not worry at all.

Sees the dead: if he is ill: It is bad for him.

Speaks with the dead: He will be joining good people, and they will like him.

Sees one of his parents dead that come to see him: He will be wealthy.

He embraces him. and kisses him: More and more wealth.

If he bit him: A misfortune will happen to him .

Sees a dead person that gave him nothing: He will receive gains.

And if he doesn't want to take from the dead what they are giving him: It will be bad for him .

If the name of the things that the dead is giving him start with (NUN)or L(LAMED): He will become poor.

If the dead stole from the dreamer: One of his relatives will die

If the dead gives the dreamer a steel tool or weapon: Where ever he goes, he will be safe and will not need to worry .

Sees that his Mother or Father died A joy will come to him;

: If they are already dead: : more so a great joy .

if they gave him nothing: greater joy.

Washed a dead person, dressed him, and carried him: He will descend from his present greatness.

He went behind the bed of the dead or comforted the mourners: A deed, he will do that will bring him closer to the Creator .

Killed a person: A miracle will happen to him.

Sees tombs: Ugly deed or occurrences.

CLOTHING

Clothing in general: Will demonstrate beauty, honor, and usefulness.

Sees black clothing: A misfortune he will endure.

Saw a cloth: Demonstrates instruction, teaching.

Find gold color cloth: It is very good for him .

Wears a new Talit (praver shawl): He will marry.

Sees himself dressed in women's clothing: She will inherit him.

Sees a woman dressed in her husband's clothing: She did a forbidden act.

A belt of silk: Greatness he will achieve.

A cloth of silver and gold: He will rise to greatness that no one could ever have imagined.

Wears expensive clothes: Greatness he will attain.

Removed expensive clothing: He will fall from greatness

Wears silk clothing: A misfortune will happen to him as punishment.

Wears red clothing: People will envy him.

Wears white clothing: Sign of good deed.

Wears black clothing: It is an omen.

All kinds of colors: Are good for the dream (but sky blue).

Sees his garments torn: His judgment has been found groundless.

The crown or turban is taken away: His luck will diminish.

Undressed: If he is sick, Good it is.

His clothes are lost: A loss is anticipated.

His clothes are burned: Suddenly he will gain a big profit and will buy a lot of clothing.

Sees himself naked: In Israel, He has no good deeds.

SUN, MOON AND STARS

The Sun: Indicates a King, Father, Master or Rabbinate.

The Moon: Government The Stars: Brothers, friends, professors or nations.

Sees a sun or moon eclipse: Indicates death of illness.

An eclipse of both: A wrath on that country.

A sun and moon in that country: It is a good prediction for that country.

The gates of heaven opened and a water course: A great wellness, will happen to that country.

Went up higher in heaven: Bad, will happen to that country.

He Sees the sky covered with very dense clouds: A great thing will be lost.

Rain falling with strength and out of season: A tragedy will happen to that town .

Thunder and lightening without rain: That generation is full of bad deeds and the Creator is angry at them.

Thunder and lightening and rain: Good will come to that country.

Sees the Book of Kings: He should expect righteousness.

Sees the Book of Issiah: He should expect wisdom.

Sees the Book of Jeremiah: He should worry about calamity, suffering and evil.

Sees the Book of Psalms: He should expect righteousness.

Sees the Book of MishIe {Book of Proverbs): He should expect wisdom.

Sees the Book of Job: He should worry about calamity,suffering and evil.

Sees the Book of Songs of Songs: He should expect righteousness.

Sees the Book of Kohelet (Ecclesiastics): He should expect wisdom.

Sees the Book of Kinot (Lamentations): He should worry about evil,calamity, and suffering.

Sees the Scroll of Esther: A miracle will happen to him.

Sees a new letter: Good predictions.

Sees a scroll of Torah that is burning. or not clothed or reads on it: It is very bad for him,he should fast even on a Sabbath.

Scroll of Torah respected: He will father a baby boy.

Sees himself praying: It is a good sign for him.

Teaching youngsters: He will be a great man to others.

If he sees himself on Rosh Hashanah and blowing the Shofar: He should fast and request pity, his misdeeds are hanging over his head.

Sees himself as a cantor in the Temple: If he is of merit; Greatness he will have, if not of merit: Shame he will have.

Who answers Amen Yehe Sheme Raba: He is guaranteed the world to come.

Reads the Keriat Shema: Is worthy that the divine presence inhabits him like in Moses blessed be his memory., only if his generation is deserving it.

Wears the Tefilim (Phalacteries): He should expect greatness.

All that I have mentioned are explanations of dream language brought to us by greatpeople. Like Joseph, son of Jacob in Egypt. He explained the Pharaoh's dreams and the dreams that included him, and his father and brothers, etc...Also the great Daniel in Babylon as well as others. As I explained before, dreams are premonitions brought to us by Intelligence from another dimension, by departed souls or other forms of Intelligence.If the dream has a bad significance, it is recommended to go to a Rabbi for "Hatava halom", with fasting and charity. If he is not Jewish, prayers of mercy, fasting and charity will do.

A WISH
A MESSAGE OF PEACE

Conclusion:A Message of Peace

It is very hard for me to comprehend anti-Semitism, when we all come from the same forefathers. And to hate Israel is to hate our forefathers. I'll proceed to explain why.

What does Israel mean? First, our forefather Jacob was called Israel. In Hebrew, Israel is said Yisrael-ישראל and it is composed of the following:

Y -י= for YACOV יעקב-& YITZHAK-יצחק (Jacob and Isaac)

S -ש= for Sarah-שרה

RA -ר= for Rachel-רחל and Rebecca- רבקה

E -א= for Abraham -אברהם(A or E are written in a single letter – Aleph-א)

L-ל = for Leah-לאה

Those are our 7 Forefathers

JERUSALEM IN HEBREW IS: YERUSHALAYIM-ירושלים

YERUSHA-ירושה = that means inheritance.

L -ל= (Leamo-לעמו) to his people

YI-י = Yisrael-ישראל

M -ם= (Meolam-מעולם) forever.

We all are creations of the Creator, Blessed be He, we all descended from ABRAHAM, and we all believe in one G-d. Division between races was made long after ABRAHAM, the picture looks like this:

Adam and Eve (all creation)

Abraham

Isaac and Rebecca

Jacob Rachel-Leah

Moses

JEWS MOSLEMS

CHRISTIANITY

The three are as a formation of an atom, or in. our case, a particle of an atom or quark. We do need each other to exist, so why kill each other. Because an atom Without a proton, neutron and electron, will cease to be one. Let's all pray for the coming of the Messiah, that the ox will eat together with the sheep Let's all together try to know our Creator better and by the same token, we should explore the infinity of our body, and universe.

the S.E.G-השאג

Printed in the United States
by Baker & Taylor Publisher Services